# WILDFIRE WARS

# WILDFIRE WARS

## Frontline Stories of BC's Worst Forest Fires

**KEITH KELLER**

HARBOUR PUBLISHING

Published by
**HARBOUR PUBLISHING**
P.O. Box 219, Madeira Park, BC Canada V0N 2H0
**www.harbourpublishing.com**

Edited by Betty Keller and Ursula Vaira
Cover and page design by Martin Nichols
Maps by Nick Murphy
Front and back cover photographs from the BC Forest Service collection
Author photograph by Heather MacLeod
Background photographs accompanying maps by Douglas Cowell
Photographs on pages 13, 81, 97, 107, 209, 259 by Bill Jackson; photograph on page 31 from BC Archives, NA-06500; photograph on page 59 from BC Archives, NA-06427; photograph on page 77 from BC Archives NA-20179; photographs on pages 125, 133, 155, 193, 231 by Douglas Cowell, BC Forest Service collection; photograph on page 169 from the BC Forest Service collection; photograph on page 251 courtesy Wayne Langlois

Printed in Canada

Harbour Publishing acknowledges the financial support of the Government of Canada through the Book Publishing Industry Development Program (BPIDP) and the Canada Council for the Arts, and the Province of British Columbia through the British Columbia Arts Council, for its publishing activities.

THE CANADA COUNCIL | LE CONSEIL DES ARTS
FOR THE ARTS | DU CANADA
SINCE 1957 | DEPUIS 1957

**National Library of Canada Cataloguing in Publication**

Keller, Keith, 1958–

  Wildfire wars

  ISBN 1-55017-278-6

  1. Forest fires—British Columbia. 2. Forest fire fighters—British
Columbia—Anecdotes. I. Title.
SD421.34.C3K44      2002 634.9′618′09711      C2002-910758-X

*For Heather—keeper of the flame*

# Contents

# Acknowledgements

In the spring of 1998 I met Jim Dunlop, who was then director of the Ministry of Forests' Protection Branch and therefore the province's chief firefighter. He was, at the time, helping lawyers prepare the ministry's defence against an anticipated lawsuit by homeowners burned out in Penticton's 1994 Garnet fire. Things must not have been going too well.

"Teaching a lawyer how to fight fire," he sighed, "is like teaching a monkey how to fly a helicopter." Before I left he gave me his personal copy of Norman Maclean's *Young Men and Fire*, and after I left he asked a lot of seasoned firefighters on his staff to pass along to me the names of people I might interview. That got me started.

I owe thanks to many other people for their contributions to this book. At the Ministry of Forests I am particularly grateful to John Parminter for his good-natured and enormously helpful responses to my many requests. In other forests ministry offices: Suzanne Barker, John Bedell, Paul Gateley, Paul Gevatkoff, Dave Hails, Jim Highsted, Tina Johnston, Diane Morrison, Lindsay Olivier, John Sloan, Wendy Stewart. In the rest of the world: Geoff Bate, Geoff Bodman, Bill Bunting, Alex Christie, Fred Derry, Vern Hopkins, Jake Jacobson, Richard Mackie, Janet Mason, the Mineault family, Colleen Minnabarriet, Marie Minnabarriet, John Noble, Janice Nousek, Tony Robinson, Shannon Scott, Lloyd Siver, Robbin Spencer, The Thursday Night Think Tank. To those going unmentioned owing to shortcomings of memory or record-keeping, thank you.

Some comments by Jack Coates in "The Final Days of Camp McKinney" were first published in "Memories of the 1931 Camp McKinney Fire," *Okanagan Historical Society Report* No. 59.

At Harbour Publishing I am indebted to Howard White for his combination of faith, insight and intransigence; to the invaluable Peter Robson for making it happen; to Betty Keller—no, she's not my mom—for being a superb editor. Thanks also to Vici Johnstone, Shane McCune and Nick Murphy.

A Grant for Professional Writers from the Canada Council for the Arts was instrumental in allowing me to complete this book. For that I am grateful.

Most of all, love and thanks to my family, Heather MacLeod, Madeline and April.

# Introduction

In the course of compiling this book I developed a theory. It goes like this: Choose three people at random, rummage through their memories and you've got better than even odds that one of them has either fought a forest fire or knows someone who has, or has had close contact with wildland fires in some memorable way. Two things will happen once you prompt people to recall their fire experiences. One, their eyes will light up. Two, they'll want to talk about it.

I was in a bookstore in Campbell River promoting a book that had nothing whatsoever to do with forest fires. A girl, perhaps eight years old, stopped to chat. She mentioned that her father worked in the forest industry. When her father joined us I asked whether he had any experience fighting fires. Jim Bufano's eyes lit up and he answered emphatically, "Yes!" Searching for a concise description of firefighting's appeal, Bufano settled on this: "It's like a war without bullets."

The lexicon of firefighting is unrepentantly military: initial attack crews, rapattacks, parattacks, direct and indirect attacks, mop-up operations and bomber aircraft all form part of the firefighting arsenal. In the place of privates, sergeants, captains and generals we have basic firefighters, crew bosses, sector bosses and fire bosses. Seeing each other in action, the firefighter and the private, the general and the fire boss would instantly recognize their counterparts.

The war metaphor explains more than the language and organization of firefighting, however. Bullets or no, the military campaigns that major wildland firefights become are extremely hazardous, and it would be improper not to recognize the risks taken on our behalf by those assigned to wage them. But the dead and injured are not the only casualties of firefighting wars described in this book. Some wounds are less evident but no less real. For example, there was the public humiliation of the emergency recruits who were summarily dismissed from the 1938 Bloedel fire. Until recently it was not unusual for firefighters to be hired on an emergency basis and sent to the fire line largely or entirely untrained. It is possible to read the term "cannon fodder" between the lines of an assistant ranger's written comments about people sent to the volatile 1960 Mil fire: "The majority of men ... were of the

# — *The Fire Triangle* —

Imagine a triangle formed by three lines connecting the follow-ing elements: fuel, oxygen, heat. A fire requires all three of these elements in order to live and grow. Remove any one of them and the fire dies. The role of the firefighter is to break one of the connections in the fire triangle. It's that simple.

Along the way, however, our hypothetical firefighter may gath-er several hundred like-minded souls and the materials required for their extended support. He may assemble vast quantities of equipment. He may summon many kinds of aircraft from across the province or even from across the country. Highways and air-space may be closed, sophisticated communication systems installed, satellite connections established, water quality moni-tored, emergency programs activated, public information dis-seminated, communities evacuated. Lines of command may be established, jurisdictions clarified, strategic permissions gained from on high. Heavy equipment operators may be sent on head-long charges down hillsides approaching the vertical. In the most volatile of conditions, more fires may be lit in the hope of pitting one element fatally against itself.

Any or all of the above may be done to achieve the same effect that you and I get by pouring water on a campfire. Only the scale and the context change.

Fuel, heat, oxygen: to kill a fire, one of them's got to go.

scruffy rubby-dub type and it gave one the impression that whoever was pick-ing them up was doing a very good job in cleaning up the undesirables from the streets of Kamloops. Many were sent out to the fire line with insufficient clothes, some in just tee shirts and sandals."

Firefighting, like war, brings out the best and worst in people. In the mid-1980s forest officer Brian Pate was sent to help fight a tough fire in BC's far north. An overhead team—the Special Forces of fire fighting—had been sent from the south to run the campaign, and in Pate's opinion they'd been running it badly. "They didn't understand the fuel types, they didn't understand the weather and they didn't understand Indians."

One day a Native man walked into camp seeking a job fighting the fire, as some Native people were already doing. This man and his family lived up the Liard River at a remote spot accessible only by boat. The O-team fire boss, Pate later learned, told the prospective firefighter that he wasn't going to hire him because he'd probably started the blaze in the first place. The man got into his boat and motored back up the river. Pate decided to make amends for that insult.

"When the fire was over I took a helicopter and loaded it with food. It was a huge camp and you've always got all kinds of food left over after a fire. I took staples—flour and sugar, all kinds of canned stuff. This guy and his family probably only got to town for supplies two or three times a year. I filled the helicopter with this stuff and flew up the river. They had a clearing with a garden, so I landed there rather than making a big racket beside the house. I just started unloading all the food, stacking it. The family was all standing there, watching me. This kid came up to me but he didn't speak English at all, so through sign language I made it understood that this food was for them. Then I just got back in the chopper and left. That was my way of making up to them for the other guy."

I thought when I began this project that I had a pretty good idea of what I would learn: that firefighting involves brothers-in-arms (gradually being joined by a few members of the sisterhood) cohesively and gallantly waging campaigns for the common good—white knights riding into battle, their lances lowered at a common enemy.

Well. I did hear tales of and from people who were exceptionally loyal, hard-working and intelligent, yet as humble as any I have ever known. I heard about demonstrations of courage, stamina, dedication and selflessness. I also heard about fights, thefts, drugs, implied incompetence, and rivalries petty and profound. I may even have met a few white knights.

# CHAPTER 1

# The Final Days of Camp McKinney

O n the evening of July 27, 1931, lookout Albert Lee, perched on the southern Interior summit of Baldy Mountain, spotted a fire burning in standing timber near the McKinney Road, a gravel track joining Oliver with Rock Creek. The area's telephone exchange was closed for the night, and it took Lee 20 minutes to communicate his discovery to the BC Forest Branch [to become the BC Forest Service in 1945]. Minutes later, assistant forest ranger Cuthbert Bodman was on his way to the fire in the company of a local rancher, Don Aulden.

In falling darkness Bodman and Aulden hiked into the bush from the end of a private ranch road, their progress along an old trail impeded by windfalls and poor light. The men discovered a 1.2-hectare fire burning briskly with considerable potential for growth. Returning to Rock Creek, Bodman arranged to have all available local men ready to tackle the blaze at first light. The sole exceptions were farmers involved in harvesting the year's hay crop, a vital rural activity from which firefighters were diverted only in the direst of circumstances.

# Last Days of Camp McKinney (McKinney Fire, 1931)

Whether conditions were dire or not was a valid question. Even measured against standards typical for the arid Okanagan–Boundary country, the summer of 1931 was remarkably hot and dry. In fact, throughout much of BC, smoke-filled valleys were more the rule than the exception. There were so many fires burning in 1931 that some ran out of room and merged their boundaries, forming larger unified blazes. It would be almost three decades before the province would record a comparable fire year.

Cuthbert Osmond Bodman was born in 1890 in the village of Barking on the Thames River and educated at Queen Elizabeth College on the island of

Guernsey, where he excelled at Greek and Latin. In 1908 his mother responded to a family breakup—her husband had run off with the governess—by immigrating to Canada with 18-year-old Cuthbert and his younger brother. Cuthbert Bodman homesteaded a quarter section at Dilke, Saskatchewan. Though he helped found that community—his ox team dragged the stone used to build the church he was married in—the farm failed. He worked briefly as a miner before going into partnership with his brother-in-law, Cecil Clarke, on a ranch located at Upper Rock Creek. A returned WW I infantryman, Clarke had qualified for financial assistance from the Soldier Settlement Board to buy a piece of property that had been settled during the southern Interior gold rush by a relative of Winston Churchill. The place included a barn, various outbuildings and a house built on a foundation of cement poured around empty whiskey bottles.

*Cuthbert Osmond Bodman. The assistant ranger had immigrated to Canada as an 18-year-old. He homesteaded in Saskatchewan and mined in BC before taking up farming and forestry-related work at Rock Creek.* Photograph courtesy the Bodman family

Bodman and Clarke raised pigs, kept milk cows, ran range cattle, maintained a small flock of sheep and grew hay and grain for livestock feed.

The year prior to the McKinney fire, while Clarke had stayed on at the ranch, Bodman had moved his family closer to Rock Creek to provide better schooling opportunities for his three sons. He had also begun working as a log scaler for sawmills on both sides of the BC–Washington border, but each summer he signed on as a Forest Branch assistant ranger, his prime responsibility being fire suppression. Bodman was a capable self-starter though he was reputed to have a volatile personality—in the church at Dilke he had been known to stand up and challenge the minister's sermon.

*The Bodman family in 1935: Irene and Cuthbert Bodman with (1 to r) sons Alan, Geoff and Phillip.* Photograph courtesy the Bodman family

Bodman led 20 recruits on the initial attack on the McKinney fire at 5:30 a.m. on July 28. Four hectares of jack pine, fir and larch forest were burning by then, the fire's intensity being additionally fuelled by a "heavy growth of moss and bug-killed trees."

Though the firefighter's quiver has been filled with many new arrows during the past seven decades, the strategic essentials remain unchanged. Any fire too large to be extinguished by a direct assault must be surrounded by a fireguard—a swath of ground scraped clear of all combustible materials—thereby preventing its continued progress. Fireguards are first established at the fire's base, then pushed along its edges, or flanks. When conditions allow, control is taken at the head—the direction the fire is advancing owing to wind or slope.

Complicating the fact that the McKinney blaze had grown to four hectares by the time of the initial attack, a strong breeze sprang up in the day's growing heat, driving the fire ahead of its mattock-wielding pursuers. Furthermore, as Bodman noted in his daily log, "There were practically three crops of jack pine, one 30 to 40 feet and moss laden, another about eight feet high . . . the last being windfalls . . . " In other words, the forest was thick with a condition known as ladder fuelling—combustibles layered in such a way that debris on the forest floor ignites young trees above, allowing flames to reach the highly flammable canopy of older trees. A fire feeding on ladder fuels can climb trees like a cat. A single tree bursting into flame in such a situation is said to be candling. Nourished by wind, a candle tree can ignite its neighbours. The cumulative wind-driven effect is a crowning fire, one capable of rolling through the canopy with astonishing speed.

As Bodman's crew struggled to hack out effective fireguards, the McKinney fire candled, then crowned. Almost inevitably, given the mounting wind, it began generating spot fires. Spotting occurs when wind-blown sparks or chunks of burning material land and ignite ahead of the main fire. Spot fires can trap crews between a primary fire and its progeny, and they can make a mockery of fireguards laboriously grubbed out of the earth. Bodman had divided his crew between the fire's north and south flanks, hoping they could push fireguards beyond the blaze, surround its head and, in firefighting argot, "pinch it off." But wind-driven sparks repeatedly spawned new fires beyond the firefighters' guards. Considering the prevailing conditions, it is likely a testament to the crew's diligence that by nightfall the fire had grown to only 40 hectares.

For two days and nights Bodman's crew—now numbering 42 men—played a cat-and-mouse game with the McKinney blaze. The fire was by now the odds-on favourite. In addition to the natural disadvantages working against him—slope, wind, heat, volatile fuels—the assistant ranger began to suspect that he was fighting a more perverse enemy. In a journal entry for August 1 he noted that one of the spot fires "was against the wind and looked suspicious." With the province in a profound economic depression the temptation to prolong a fire or light new ones was more than some people could resist, the hourly wage of 25 cents notwithstanding. Investigating the possibility that he had an arsonist among his firefighters, Bodman spent much of the day disappearing into the forest and reappearing randomly, "coming out on the crew from unexpected places." He discovered nothing to confirm his suspicions. Owing to intense heat his north and south flank crews were by now forced to build their control lines well back from the fire. The two crews were separated by 800 metres, much of the space between them being occupied by "a regular furnace."

Though he continued to suspect that someone was lighting fires outside the lines, Bodman took time to note that, "The transient members of the crew seemed an exceptionally hard-working bunch." His transient workers were men drawn from a nearby federal unemployment relief camp, one of many such facilities established to house and occupy single unemployed men during the Great Depression. Bodman's son Geoff, seven years old when the McKinney fire occurred, had first-hand contact with some of the camp recruits, whom his father had at times invited into his home to play bridge.

> Geoff Bodman: After the strikes in Vancouver, when the unemployed sat down in the post office and things came to a head, the government said, "These men have got to go somewhere." That's when the Department of National Defence started building emergency airports and roads all over the country. There was a camp of 75 men there in the [Kettle] valley, men from all over Canada. Dad needed men for the fire one day, so he went to that camp and hired a bunch of them.

By August 3, suspicious spot fires continued to plague Bodman, who had become convinced that he was dealing with fire-line sabotage. Whether he felt that he'd caught the culprit remains unrecorded, but he removed one man from the crew for being "unsatisfactory," delivered him to Rock Creek, then drove to nearby Westbridge, where a new blaze had broken out. While there, he glanced westward and, noting the growing smoke column, predicted that "a blow-up on the McKinney fire is imminent." Flames blew wild that afternoon. Rushing back to Rock Creek, Bodman and a second Forest Branch official began making preparations for an expanded fight. Up to this time the McKinney had remained a wildland fire, burning in forest cover only. It was now becoming—though the term had yet to be coined—an "interface" fire, one which also threatened homes and other developments. This was confirmed when Bodman received news that "certain people were in danger" north of Rock Creek. One of the threatened ranches was his own and one of the people in danger was his partner and brother-in-law, Cecil Clarke. Bodman later learned that when gale-force winds blew the McKinney out of control, Clarke had piled a few essentials into his Model T touring car and made a last-minute escape. As he fled, one of the embers raining around him ignited a mattress crammed into the car's back seat creating a mobile spot fire that he was forced to extinguish. He then took refuge at the home of rancher William Hatton. Bodman also visited the Hatton ranch that day, though he made no reference to encountering Clarke there. He did meet Gwen Hatton, William's wife, and suggested that she evacuate to safer ground. Gwen refused, stating that she would stay to fight spot fires around the house and "to stuff some belongings into trucks in case of a quick getaway." The Hattons did take the precautionary step of evacuating their five-year-old daughter Margaret to safety in the Kettle Valley, sending her away with a neighbouring rancher who evacuated 11 children in his Model T. Margaret Hatton—now Margaret Henley—remembered seeing flaming pieces of bark launched by the main fire front across the Rock Creek Canyon land in the family's extensive hay and grain fields. She also recalled seeing her mother—"a little bit of a thing and a hard worker"—coming into the house with her face blackened by smoke from battling the resulting flames. A Hatton family friend arrived at the ranch to offer assistance that day. Uncharacteristically, he arrived on foot, explaining that he had begun the journey on horseback but had been forced to tie the animal up and continue on shank's mare

*The Bodman-Clarke ranch house in 1920. The house and all other buildings on the ranch were destroyed in the McKinney's wild August 3, 1931, blow-up.*
Photograph courtesy the Bodman family

owing to the shotgun blast of sand being driven in his face by the tremendous wind. Nearby, other ranchers worked desperately to plow fireguards around their properties in the hope that bare soil would check advancing flames. Some of them succeeded.

Leaving the ranch, Bodman attempted to cross the Rock Creek Canyon but was driven back by fierce heat. He continued his patrols until well into the morning of August 4, slept briefly, then returned to the fire. By then he had almost certainly learned that the Bodman-Clarke ranch had been burned out the previous day. Of the house, barn, sheds, fences and machinery, only one outbuilding remained. Five cattle had been killed. Spot fires burned over a wide area, but determining their location was difficult, as smoke often reduced visibility to near zero.

# — *Boiled Fish* —

A half-century after the McKinney fire, rancher William Hatton published a memoir in which he described August 3, 1931—the day of the blaze's most destructive blow-up—as "one day never to be forgotten." The Rock Creek district, he wrote, "was well wooded, with superb scenery in those days, but it was soon transformed into an inferno of swirling flame and thunderous roaring. Horses, cattle, deer and all living things in the path of the fire met death, including fish in the creek."

The Johnstone Creek area west of Rock Creek was Crown range land where several ranchers ran stock. Cattle in that area were "wild and scary," Hatton wrote, and getting them rounded up and driven to safety was a near impossibility. Some animals were later found alive; others died when they stampeded down into the Rock Creek canyon. Many injured cattle had to be shot.

Geoff Bate, born two years after the McKinney fire, was raised on a ranch north of Rock Creek. As did Geoff Bodman, Bate went on to make a career in the BC Forest Service. As a kid he used to ride an old horse named Buck. A survivor of the McKinney fire, Buck had terrible scars on his shoulders and neck—scars left by chunks of burning material that fell on him as Bate's father attempted to roust cattle out of Johnstone Creek during the August 3 disaster.

In 1930, when the Bodman family moved to Rock Creek to be closer to school, they had presented to Cecil Clarke a cat whose job was to reduce an overabundance of mice on the Bodman-Clarke ranch. Clarke had to leave the cat behind when flames drove him from the ranch, and her fate was unknown until ten days later when she turned up at the Bodman home. She was thin and nervous, and had the fur singed from one side of her body. She remained a steadfast mouser for years after.

On August 6, Bodman's embattled force was bolstered by reinforcements recruited from Oliver and Penticton. They established a headquarters at the gold mining ghost town of Camp McKinney and for accommodation took over a building being squatted in by a prospector known as Gunsight Grant. Bodman met with the foreman of the new arrivals, the two men agreeing that their priorities should be to defend a 60,000-volt West Kootenay power line running through the area—the line providing electricity to the Okanagan Valley—and to protect valuable timber stands along a Rock Creek tributary known as Jolly Creek.

One of the Oliver contingent was a 22-year-old man by the name of Jack Coates. An aspiring orchardist, Coates had just purchased a parcel of raw land at Oliver and, needing money to develop the property, had hired on to fight the McKinney fire. He and his crewmates had been working out of Camp McKinney for three days when Bodman's fire sense told him that another blow-up was inevitable. Maintaining a constant truck patrol along the power line, Bodman transferred men "to and fro to swat the [spot fires] that were increasing in number." When flames cut off the road, he could see only one recourse: "to pile the more valuable stuff from the Cariboo [Camp McKinney] fire camp, park the truck and get the men out to safety as the fire then was only 200 yards from the camp."

*Retired orchardist Jack Coates at 92. He survived a blow-up of the McKinney fire by hiding in a mining tunnel while crowning flames roared overhead.*
Photograph courtesy Jack Coates

Jack Coates, now 92 and retired from an orcharding career on land developed in part with his paycheque from the McKinney fire, recalled the evacuation that Bodman prepared.

> Coates: One morning we were called together and told that the fire was out of control and that we would have to evacuate the camp immediately. There wasn't even time to collect our personal belongings. Bridges were burned out and the road was cut off, so we walked the power line to Rock Creek, about 12 miles, then were taken by truck to our next camp. The fire totally destroyed what was left of the town of McKinney as well as our provisions, tools and two West Kootenay Power company trucks.

Coates and his companions established a new camp north of Rock Creek, near a ranch owned by bachelor Billy Munch, "a character and quite a rugged individual," as Coates remembered him. Bodman's crews helped Munch take whatever fireproofing precautions they could, then switched their focus to a sawmill operated by the Rock Creek Land Company. Strategic backfiring held off the oncoming flames until two o'clock in the morning, when a shift of wind defeated their efforts; the mill was absorbed into the inferno.

The Jolly Creek sector remained an ongoing struggle as did the Rock Creek Canyon, where at least two close brushes occurred. Flames racing down Rock Creek took a fire crew by surprise one evening as they sat down to supper. The camp cook evacuated a carload of men just before the camp and most personal belongings burned. "They hardly succeeded in making their way out," the *Penticton Herald* reported, "as the smoke was so dense that it was almost impossible to see a few feet ahead of them."

Jack Coates had his own escape from flames when he was sent to the aid of a fire crew in danger of becoming trapped.

> Coates: At one point, with the fire becoming more and more threatening, Mr. Bodman got worried about a group of men working in the Rock Creek Canyon. He asked three of us, Johnny Haggart, Shorty Graves and me, to go in and bring them out. Fortunately, my companions had been placer mining and knew the area. There was a strong wind blowing up the creek and it sounded as though a freight train was coming up the canyon. And coming fast.

We just made it to an abandoned mining tunnel as the fire crowned over us. Fire was roaring through the treetops, with burning embers flying hundreds of feet ahead and igniting everything in their path. As the fire subsided, we were able to pick our way back to camp. As it turned out, the other crew had been aware of the danger they were in and had gotten out safely.

It is almost certainly true that the McKinney blaze at its worst was unstoppable by any firefighting force, in 1931 or today. On the other hand, it is possible that a larger initial response may have allowed the fire to be controlled before it got out of hand, as several fire victims vocally suggested when they pressed claims for compensation against the provincial government. Unfortunately, BC's high-intensity fire season of 1931 coincided with a depression that strained government services. As well as reducing staff dedicated to fire detection and suppression, the Forest Branch had issued a directive stating that "where a fire was not caught and controlled in its incipient stage, field officers must give consideration to the values protected, and that intensive fighting should not be undertaken to protect stands of low value belonging to the Crown." Bodman made no written reference to this directive's influence on the McKinney campaign, but it is not difficult to imagine that he would have appreciated having a Forest Branch theorist standing beside him at some point during the fight, interpreting that policy as it pertained to the cataclysm rolling across the timber—some of it high value, some of it low—and ranches of the Bridesville–Rock Creek countryside.

Three weeks before the McKinney fire ignited, a blaze had been reported near Inkaneep Creek, southeast of Oliver. The Forks fire, as it became known, burned over much of the Anarchist Mountain ranching country spreading east above Osoyoos Lake. Edward Lacey was one of the people leading the Forks firefight. In addition to being a farmer and government road foreman, Lacey was an honorary fire warden, a citizen empowered to fight wildfires breaking out in his vicinity. "He had about 90 men working for him," recalled son Ed, himself a former fire warden. "There were no four-wheel-drive trucks or anything. He had to hike them guys out through the goddamn brush from camp every morning. The camp was on Kehoe's Ranch, up on Anarchist Mountain."

By August 12, the Forks and McKinney fires merged with each other and with another fire, the Smith, which Forest Branch supervisor J.W. McCluskey had discovered on August 8 while working on the McKinney blaze. McCluskey had hastily gathered up 30 firefighters and transferred them to the Smith. Only half of that fire had been guarded by the following morning when a strong wind blew it out of control up Baldy Mountain, nearly trapping the crew. One report indicated that the fire erupted upslope at a rate of five kilometres in 20 minutes. It eventually burned out in alpine scrub just short of the fire lookout from which Albert Lee had discovered the McKinney blaze two weeks earlier.

Cuthbert Bodman's journal became increasingly abbreviated as the fight against the combined fires wore on, providing no detail as to what it may have looked and felt like when, for example, on August 19, "the James Lake sector blew up and we had to retire to another ridge."

> Coates: We were starting to get frosty nights by this time. It was cold getting out of the sleeping bag when the breakfast bell rang. Our cook was Louis Lasalle, a Frenchman. As he rang the breakfast bell he'd call out, "Birdies are singing, sun is shining, daylight in the swamp, come and get it!"

When discrete sections of fireguard were finally connected and held—using nothing but hand tools, crews had built 40 kilometres of control lines on the McKinney alone—it became apparent that Bodman's men were being rewarded for their dogged toil. Firefighters were gradually laid off until, on September 7, the fight was declared over. The three combined fires covered more than 30,000 hectares of land, with the McKinney contributing the majority of that area.

An investigation concluded that the McKinney fire was arson; local scuttlebutt suggested that it was lit out of spite by one farmer who held a grudge against his neighbour. Though also ascribed to arson, the Smith apparently was lit by a farmer burning off land for agricultural purposes; in the prevailing burning conditions, the distinction between land clearing and incendiarism would have been subtle indeed. The Forks fire was officially declared the result of careless smoking, but arson played a part in it, too. In mid-August, forest supervisor McCluskey was making a morning inspection near

Kehoe's ranch on Anarchist Mountain when he surprised a man hunched over a small fire on a dirt trail. Police eventually arrested and charged with arson a man living in a hobo jungle near the Penticton rail yard, the culprit being described in the *Penticton Herald* as "a transient, [who] speaks with a broken accent and was employed as a firefighter." Arnold Otto Andrew was sentenced to 23 months of hard labour in Oakalla Prison, despite having presented what a *Herald* reporter described as "a stirring appeal for himself, insisting that the court was sending an innocent man to jail." The sentence seemed particularly harsh considering that superintendent McCluskey, the man who laid the incendiarism charge, testified that Andrew's fire appeared to have been in no danger of spreading from the trail where it was built. Arson was, at the time, the cause of many fires burning in the region and the stiff punishment may have been intended as a statement of judicial intolerance of that crime.

Amazingly, considering the almost complete lack of fire-line communications and transportation, no major injuries were recorded in the Forks–McKinney–Smith conflagration.

> Lacey: They never lost any men in that fire but they had some close calls. My father was missing one man, so he took a couple of men and a saddle horse and went looking. They found him running around in circles where the fire had been. He wasn't hurt very bad but he had some burns. He'd been right in it. They got him out and got him to a hospital.
>
> That fire crowned about three times and when it crowns you're in real trouble. You don't know where the wind is going to take that fire. Once it gets up in the trees it creates its own wind and anything can happen. When the fire crowned there was guys that had the shit scared right out of them.
>
> I'll tell you how bad that fire was. The smoke filled this [Okanagan] valley so thick the cantaloupes never got any net on them that year. Cantaloupes need sun to develop that net, but the smoke was too thick. That's the only time I ever saw that happen.

When it came time for paycheques to be distributed, some firefighters received infuriating news. Men who had been drawn from the federal relief camp—the unemployed "transients" whom Bodman described in his journal as working as hard as, and in some cases harder than, the local men on his crew—were told that the government couldn't afford their wages. Having toiled for weeks on sweltering fire lines, slept in makeshift camps and been chased by wildcat flames, they did not accept the news well. Some decided to vent their anger at the man who'd recruited them—Cuthbert Bodman. Two men from the camp, who had gotten to know the assistant ranger over hands of bridge at the Bodman home, informed the provincial police that a group of disgruntled firefighters planned to stone his house. Police attended and the retribution was averted, but the prospect of being attacked in his home made a lasting impression on seven-year-old Geoff Bodman, who believes that the men were never paid for their work.

The McKinney fire became one of the most expensive suppression efforts the province had experienced, even without the additional costs of the Smith and Forks blazes. Forest supervisor J.W. McCluskey explained to superiors that he had authorized the expenditure of more than $15,000 "for the reason that it threatened valuable timber and young growth and the settlements of Bridesville, Rock Creek and Anarchist Mountain." Such an expenditure was cold comfort for settlers who lost property to the McKinney conflagration. They held the feet of Forest Branch bureaucrats to political flames even before fires on the ground had been extinguished.

Rancher William Hatton led the charge on August 11 when he demanded in a letter to Ernest Manning, BC's acting chief forester, "that this fire be handled in a more competent manner or else the whole country will be burned out. When the fire started, it was a small affair and there is no excuse for permitting it to get to its present proportions, as everyone around here realizes."

There is no disputing Hatton's contention that the McKinney fire started out small—all fires do. The validity of his second charge—that there was no excusing its magnitude—is impossible to verify in distant hindsight. Forest fires that destroy homes and other improvements are inevitably followed by charges of incompetence against the people responsible for controlling them. People suffering losses to natural disasters react instinctively, though not always justly, with condemnation. Forest Branch employee George Melrose, in justifying actions taken on the McKinney fire, told Manning that

Hatton presumed that "when we assumed responsibility for fighting [the fire] we naturally assumed responsibility for it getting away."

Burned-out Rock Creek settlers certainly assumed that the Forest Branch was responsible for the McKinney's destructive rampage. "This fire," they petitioned, "starting up so close to a settlement and therefore doubly dangerous, was ineffectually fought with insufficient men in the first place when it covered a few acres and could have been extinguished, not merely brought under temporary control." The co-author of this petition—along with the indefatigable Hatton—was Cuthbert Bodman's brother-in-law and land partner, Cecil Clarke. Another signatory was Don Aulden, the rancher who had accompanied Bodman on his initial reconnaissance of the fire. The devastation to Aulden's homestead was both telling and touching in his simple written enumeration: "Lost all."

When George Melrose met with Hatton in September of 1931, he explained that the Forest Branch's mandate required it to protect Crown timber, not private property. Forest protection staff—funded primarily by a tax on timber holders—"cannot be taken in any sense as a rural fire brigade for the protection of private property."

## — *Hobo Motivation* —

There was no shortage of prospective fire-line help available to forest rangers during the Great Depression. The problem was a shortage of money to pay them. During part of the Dirty Thirties even the 25-cents-per-hour wage was eliminated in favour of providing firefighters with food only. Forest officers didn't fare much better. Bill Noble became a ranger with the BC Forest Branch in 1919, working out of Birch Island, near Clearwater. During the early 1930s he was given twice the work and less pay to do it on, son John Noble recalled.

> Noble: The depression came on and the government doubled up ranger districts. One junior ranger from each of two adjacent districts was demoted to an assistant ranger. So my

dad took over from just north of Little Fort, about 60 miles north of Kamloops, pretty well through to the Alberta border. One man. I can remember my father coming home during that time and telling my mother, "Well, we got another 10-percent pay cut."

During the Depression the forest officers would go down and pull the unemployed off the trains while they were riding the rods. You took whoever you could get. At that time the pay was two-bits an hour and your board.

Dad had one fire in the McMurphy area—I think it was called the Lost Creek fire. It was threatening poles that pole cutters had been cutting all winter. So the pole cutters were all on the fire line, but there weren't enough of them. So they got these unemployed crews off the trains and whatnot. These pole cutters discovered that the hobos were just throwing their tools into the fire. So they began spacing them—a pole cutter, a hobo, a pole cutter, a hobo, and so on. One hobo threw a tool into the fire. Two pole cutters picked him up and threw *him* in. This guy grabbed his tool and jumped back out again. That was the last tool thrown into the fire. These pole cutters had worked all winter and that was their money laying there, less than a quarter mile from this fire. And they weren't going to let anybody slack off.

Melrose and Hatton apparently left that meeting with greatly differing impressions. Melrose subsequently reported to Manning that, having gone over "the whole question" with Hatton, "he seemed to be fairly well satisfied that as far as the Forest Branch is concerned we had done all that was possible..." Hatton penned his own salvo to the chief forester, describing how Melrose "attempted to justify the manner in which the fire was fought, which, on my part, was disputed." The situation required an investigation, Hatton concluded, "not a smoothing iron, as Mr. Melrose has attempted to apply." However, no further investigation was conducted into the McKinney fire, and settlers received no compensation for their losses.

Geoff Bodman recalled that the McKinney disaster "caused a lot of strain on family relations," with his father and uncle on opposite sides of the dispute. The Bodman boys—Geoff and his two brothers—were anxious to return to the burned-out Bodman-Clarke ranch to learn the fate of creatures they had come to know while living there, but their father refused to allow them to make the trip. He may or may not have known that Clarke was, with William Hatton, co-author of the petition charging that the McKinney fire was "ineffectually fought," but he was painfully aware of what Geoff describes as "the animosity which built up within our family over the handling of the fire suppression and the loss of material things"—animosity also harboured by former ranching neighbours. It was

*William Hatton, photographed in 1971. He accused the BC Forest Branch of mishandling the McKinney fire and demanded an investigation rather than the "smoothing iron" that he felt the province was applying to cover its alleged mistakes.* Photograph courtesy Margaret Henley

at least a year before the Bodman boys were allowed to accompany their father to the ranch and then it was only to arrange the sale of the Bodman share of a steam engine, threshing outfit and grain separator that several ranching families owned in common. Disposing of it was necessitated by Cecil Clarke's decision, taken in the McKinney fire's wake, to abandon ranching. He drove to Victoria in his Model T and spent the remainder of his working life as a landscape gardener in that city.

Cuthbert Bodman remained in Rock Creek until 1937, at which point he too moved to Vancouver Island where, among other assignments, he developed provincial parks at Qualicum Falls and Englishman River Falls.

# CHAPTER 2

# The Great Fire

The great Vancouver Island fire—you will hear it referred to historically—is already a provincial calamity.

Within hours it may reach proportions of a national disaster.

It's a picture too big for any one man to paint from any one point.

It has travelled—as this is written—some 40 miles, and some 200 square miles lie smouldering in its wake.

It has cost thousands in wages and equipment.

It has cost millions in timber and property.

It is now gnawing into the standing timber—eating away the future prosperity of British Columbia.

> —Harold (Torchy) Anderson, writing in the
> *Vancouver Daily Province*, July 23, 1938

There's no question about it, Harold Anderson had a way with words and emotions. And who better to cover a fire story than a man called Torchy?

# The Great Fire (Bloedel Fire, 1938)

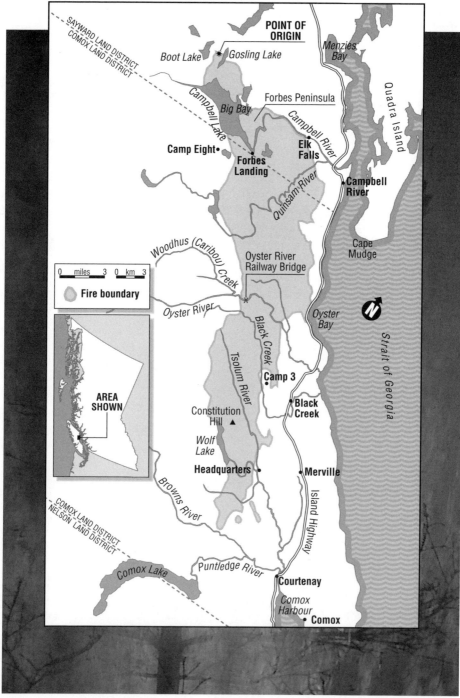

In July 1938 the *Daily Province* could have gathered vivid fire-line reports from any of several trouble spots in southwestern BC: Sechelt Inlet, the Chilliwack River Valley, Saltspring and Gabriola islands, among others. In spite of protracted drought, the BC Forest Branch had not taken the precaution of imposing a coastal logging closure. Struggling under the burden of Depression unemployment, the government of Premier T. Dufferin Pattullo seems to have preferred the risk of wildfire to the prospect of having thousands more restless men idling on Vancouver's streets.

*Smoke cloud from a blow-up on the Bloedel fire. Owing to primitive communication and poor transportation, this sight was often the only warning firefighters had that a front was heading their way.* BC Archives photograph NA-06427

On the evening of Friday, July 15, the *Daily Province*'s nightly radio news broadcast included an appeal, on behalf of the Forestry Branch, for men to help quell a major blaze near Campbell River. By the time the CPR steamer *Princess Norah* departed Burrard Inlet for Vancouver Island that night, 211 instant recruits had been shunted aboard. Many were members of the Relief Project Workers' Union, dubbed "tin canners" for having pelted Vancouver City Hall with tinned food to protest government inaction on their economic plight. These were men who had been told, in some cases for years, that their strength and skills were valueless, that they were a drain on the society that grudgingly subsidized their existence. And then, within hours, they were being transformed into foot soldiers answering that same society's call to arms.

Among the crowd was Harold Crummer, a 27-year-old from Indian Head, Saskatchewan, who, like the other men in his company, considered even the pathetic firefighting wages being offered an improvement on the indignity of protracted unemployment. He had moved to Vancouver in search of a job and instead found the mayor reading the Riot Act and mounted city policemen chasing unemployed protesters up Granville Street. Crummer had never fought fire; before long he'd be hoping that he'd never fight one again. But for now he and the other recruits, the newspaper, Torchy Anderson—everyone—had a cause.

At that time logging in the Vancouver Forest District was creating 17,000 hectares of cut-over land per year, with the result that prodigious quantities of slash were accumulating—limbs and tops as well as whole trees felled but rejected for being of less than perfect quality. Some companies had voluntary slash-burning programs in place, but even for them a substantial backlog of material existed, as did slash from the current season. To deal with the resulting fire hazard, BC's legislature had recently amended the Forest Act, requiring slash disposal at least once per year. But while the new law had come into effect on January 1, 1938, burning was not to take place until fall, when cooler, damper conditions would reduce the risk of such fires escaping and running wild across the countryside. In the meantime, slash continued to pile up and cure in the summer heat.

The emergency to which Crummer and his fellow recruits responded had declared itself 10 days earlier in one of the vast clear-cuts that were replacing

*Controlled slash burning by Bloedel, Stewart & Welch in 1929. The 1938 logging season was the first time that slash disposal was mandatory in BC, though some companies already conducted their own burning programs.*
BC Archives photograph NA-05746

primeval forest along Vancouver Island's east coast, the superb even-aged Douglas fir stands having sprung from the ashes of a massive fire that had burned out the coastal plain between Campbell River and the Comox Valley in 1668. The clear-cut that was on fire was located northwest of Campbell River, between Gosling and Boot lakes, on a timber claim operated by Bloedel, Stewart & Welch (BS&W), forerunner of the forestry giant MacMillan Bloedel. There, in preparation for shipment out of the woods by railroad, great quantities of logs had been gathered into piles known as cold decks.

On July 5 Bloedel logging crews had completed their work and returned to Camp Four at Menzies Bay. Late that afternoon a spark chaser—an employee detailed to watch for and suppress fires arising from company operations—was also heading for camp when he noticed smoke rising from one of those mountainous cold decks. Upon receiving the spark chaser's urgent message, the company dispatched 185 loggers and two pump-equipped locomotives. This force arrived on the scene within two hours and discovered that the fire had already engulfed a second log pile and was burning over two hectares of ground. Then the light westerly breeze shifted northwesterly and freshened, launching sparks that touched off secondary spot fires and advancing the main fire's head in leaps and bounds. By morning flames had burned southward to the shore of Lower Campbell Lake.

As it had started on a BS&W timber claim, fire suppression was the company's responsibility and indications are that they tackled the blaze diligently. A light rain began late in the afternoon of July 7 and continued intermittently over the next four days, giving loggers hope that they might soon exchange hoses and mattocks for the more familiar tools of their trade.

On July 12 the sky cleared, temperatures rebounded and the northwest wind resumed. By midafternoon the next day the Elk Falls lookout man reported flames advancing aggressively. Travelling up Lower Campbell Lake by boat, forest ranger Charlie Langstroth found fire-generated whirlwinds

*A Bloedel, Stewart & Welch locomotive pulls a trainload of logs toward Menzies Bay, circa 1926. Extensive railway logging in the 1920s and 1930s helped set the stage for the Bloedel fire.* H.W. Roozeboom photograph, Vancouver Public Library Special Collections

dancing like flaming dervishes along a ridge just north of the lake. Langstroth's overriding concern was that sparks might catch a foothold on the nearby Forbes Peninsula, a triangle of land bounded on one side by Lower Campbell Lake and on the other by the Campbell River. Loaded with slash, the peninsula could act as a springboard propelling wind-driven sparks into the almost unbroken mosaic of new and regenerating clear-cuts stretching down-island. The only thing separating oncoming flames from the peninsula was an arm of Lower Campbell Lake known as Big Bay.

However, arriving back at his Campbell River headquarters, Langstroth received a phone call informing him that spot fires were already burning on the peninsula. With Forest Branch supervisor Bert Conway, Langstroth raced via Elk River Timber (ERT) speeder to a point overlooking the critical area. Though flames did not yet have an aggressive hold, both men realized what could happen if the peninsula took off seriously the next day—"going up in the heat of tomorrow," as Langstroth put it. They decided the best strategy would be to pre-emptively torch as much of the peninsula's slash as they could before that happened.

At midnight, Langstroth and a crew of five began lighting a kilometre-long burn-off. Conditions were less than ideal, however, and it soon became clear that the tactic would not consume enough fuel to significantly impede the fire's advance. Anticipating the worst, pumps and hoses were rushed to the peninsula's southern tip, across from the tiny settlement of Forbes Landing where, at the lake's outlet, James and Elizabeth Forbes's hotel had been catering to visiting sportsmen for 24 years. At daylight on July 15, when Langstroth discovered that the fire had jumped the Campbell River below Forbes Landing, he detailed a crew of men to attack the fresh blaze. They were briefly successful. Then the fire spotted across Lower Campbell Lake into steep slash above Forbes Landing.

With the situation deteriorating and logging crews already working flat out, the Forest Branch made its urgent request for Vancouver reinforcements. The recruited tin canners, having travelled all night, arrived at Campbell River on the morning of July 16. Harold Crummer hadn't slept on the water crossing. Unfamiliar with ocean travel, he had stayed awake, monitoring the sound of the *Princess Norah*'s engines, certain that the steamer would founder before reaching Nanaimo. Driven to Campbell River, Crummer and his fellow recruits were dispersed to help defend two threatened logging operations:

*James and Elizabeth Forbes in the main room of their popular lodge, which catered primarily to anglers.* Campbell River Museum photograph #7478

*The Forbes Landing Lodge at Lower Campbell Lake. The fight to save Forbes Landing would become one of the Bloedel fire story's most dramatic and controversial chapters.* Campbell River Museum photograph #10445

ERT's Camp Eight and the headquarters of T.J. Brown, located near Forbes Landing. The Brown situation was quickly resolved: the next day the camp, its trucks and associated logging equipment went up in smoke along with two million board feet of logs.

The Camp Eight battle became a drawn-out affair. Crummer recalled the incredibly difficult conditions facing the Vancouver recruits and the 500 loggers they worked beside as they grubbed out fire lines with shovels and mattocks.

> Crummer: Breathing in the smoke all day, it got so that you had no appetite. That smoke made you sick, breathing that in. It was a long time before I got back to eating right. We were working real close to the fire sometimes. When I came back to Vancouver people said to me, "What happened to your eyelashes and your eyebrows?" They'd been singed off. It was unbelievable that we could stand that heat.

A few Vancouver recruits were detailed to join firefighters attempting to save the Forbes Landing Hotel, now threatened by flames advancing from two sides. Writing from the isolated lodge, Torchy Anderson was having no trouble generating gripping dispatches for *Province* readers. Describing the vast drama unfolding around him, Anderson wrote: "Hundreds of men, scores of pumps, fifty miles of hose, snorting caterpillar bulldozers, axe and shovel crews—every available means of modern forest firefighting—is pitted against the red enemy. It has made a blazing hell-pot north of this little hotel, and in the slash to the southeast, a series of small infernos where old snags blaze like candles on the devil's birthday cake."

The snags Anderson so graphically described were the spire-like trunks of dead trees. Though new legislation would require snags to be felled in conjunction with logging, for now they stood by the thousands, and Anderson's analogy fit—the devil's candles were giving firefighters a hell of a time. Fireguards could stop advancing ground fires, but blazing snags released embers which carried on the northwest wind like clouds of incendiary pollen. Blowing over firefighters' heads into unguarded fuels, the embers blossomed anew, forcing crews to fall back repeatedly to re-establish their control lines.

"28 Flee Flames As Fire Sweeps Forbes Landing" announced a *Province* headline on July 19. In spite of the best efforts of 25 men "fighting heroically to the last," Torchy Anderson wrote, the first lodge buildings "crackled up before flames hard-driven by an increasing northwest wind at two o'clock, an hour after Mrs. James Forbes and her two daughters had evacuated." The hotel had been thought safe until an hour earlier, the journalist noted. "Then the dreaded northwest wind, which had been gathering momentum since morning, strengthened markedly, rolling the fire along with a speed and impetus that defied desperate efforts of firefighters, who have been guarding the landing since the blaze swept out of Menzies Bay area slash."

But Anderson's story had an element of prescience about it—he was announcing the flaming demise of the Forbes Landing Hotel nearly 48 hours before it actually burned. Fire crews preparing to retreat to Camp Eight had reported its imminent destruction via two-way radio; learning of the dispatch, Anderson had reported the hotel's destruction as fact. The next day he

*Exhausted firefighters catch a few hours' sleep near the Campbell River bridge. They had been driven from their camp by rampaging flames the previous night.*
BC Archives photograph NA-06455

*Ash and twisted bed frames lie where the Forbes Landing Lodge had stood. One account blames the destruction of the lodge on the sabotage of a critical water pump.* BC Archives photograph NA-06448

explained in the *Daily Province* that a last-minute wind shift had occurred and the place was saved by firefighters pumping water up from the lake to direct a stream of water on it. They ensured their own safety that night by sleeping on one of the resort's floats.

When the destruction of Forbes Landing was prematurely reported, a young woman burst into the Campbell River Forest Branch office demanding confirmation or denial of the rumour that not only had James Forbes been incinerated with his lodge but 150 firefighters were also missing, her husband among them. Though both rumours were false, Anderson's original report did become accurate the next day when a malfunction silenced the pump that had been preventing sparks from igniting the lodge. Within hours, more than two decades of investment was reduced to blackened ruins. Courtenay logger Walter Edwards remembered fighting fire at Forbes Landing on July 20, then hanging some water-soaked clothes to dry on a lilac bush outside the hotel before walking over tree-clogged roads to spend the night at Camp Eight. Returning the next morning, he found rubble where the hotel had stood but his dried clothes still hung from the lilac bush nearby.

As dramatic as the Forbes Landing saga was, it was by no means the only significant battle taking place along Bloedel fire lines. To the east of the Forbes's ravaged hotel lay the settlements of Campbellton and Campbell River and, between them and Lower Campbell Lake, a stand of timber that was destined to become Elk Falls Provincial Park. Treasured for its magnificent trees and spectacular falls, the park, as it was already known, became one of the points at which firefighters dug in for a major challenge—preventing the fire's continued eastward spread toward settled areas. However, a fire camp located near the park had barely been established when nighttime winds drove flames across the Elk Falls road, cutting crews off from the outside world. With the camp in "a precarious situation," as Forest Branch supervisor Bert Conway described it, a new road was hurriedly bulldozed in as an escape route. The following night, despite the fact that pumps had been installed to continually hose down two nearby bridges, a major flare-up destroyed both structures and forced firefighters into a hasty retreat. Saving the park, ranger Langstroth noted in his journal, had suddenly become "a doubtful proposition."

With the fire camp in danger of being burned over, crews threw together a footbridge over the river just upstream from Elk Falls. During the day of July 19, while one shift of men fought the oncoming flames, a second prepared for evacuation to the river's opposite bank. Portable camp equipment and food stores were carried to safety across the makeshift bridge that night. Heavier equipment was buried in place. Fire crews drained the fuel tanks of several vehicles caught in the trap—ranger Langstroth's car, a Forest Branch truck, a privately owned truck and a bulldozer—then shoved them down the bank into the river.

While coastal residents anxiously monitored the pall of smoke marking the Bloedel fire's advancing eastern edge, and hundreds of loggers and hired men fought to save Camp Eight—one of the coast's largest logging operations—a seemingly unstoppable front burned southward, eventually entering Comox Logging and Railway Company claims. For three days and nights the blaze advanced across Comox Logging land, all but surrounding the enclave of Camp Three, home to at least 300 employees and family members. The danger of being burned out was very real to these people: Camp Three had been destroyed by fire in 1925 and partially damaged twice before that. Company crews were joined by "every able-bodied man who can get away from Courtenay," Anderson wrote. As the fire's main body continued southward on one side of the camp, an ominous arm of flame was creeping around the other. Though the camp would eventually be situated within an ever-tightening fiery embrace, tenacious firefighting prevented flames from completing the lethal circle.

South of Camp Three was Headquarters, Comox Logging's centre of operations, site of its railway machine shop and home to company managers and their families. Situated in a dense tract of young, highly flammable second-growth fir, Headquarters stood directly in harm's way. Between Headquarters and the fire stood Constitution Hill, a geological loaf aligned north-south in the band of coastal lowlands. Of great interest to everyone bracing for the arrival of flames was the path the fire would take upon reaching the hill's precipitous northern slope. Burning to westward, it would endanger Comox Logging cold decks near Wolf Lake and threaten the standing forests of Forbidden Plateau. Shifting eastward, the fire would imperil Headquarters and farms northwest of Courtenay.

When the red flood struck on July 22, it parted around Constitution Hill like a rogue wave slamming a ship's bow, but it also exploded up the hill's steep face. Until now the fire had burned almost exclusively in logged-over land and young regrowth. Constitution Hill, however, remained covered in forest. Given the combination of an upward slope—an element favourable to fire spread—and parched forest, a crowning fire developed. Great clouds of ash and debris blocked the sun, forcing drivers on the Island Highway to turn on their headlights in midafternoon. A rain of embers fell in Courtenay, where the event was dubbed Black Friday. While a constant northwest wind ensured that the town of Campbell River remained smoke-free, the volume of material sucked into the atmosphere was such that people on the prairies reportedly swept ash off their cars.

As flames—audible from five kilometres away—roared over and around Constitution Hill, a crucial battle line was developed to save Headquarters. Visiting the largely abandoned settlement, the omnipresent Anderson found "an unnatural stillness" as fire crews waited for whatever came first—spot fires or the main flame front. "The cosy gardens with their sweet peas, roses, banked rows of standing corn looked incongruous," he wrote. "Hoses lined the street and emergency ladders stood at every telephone pole. Water poured down on the roofs of the buildings. Ashes and debris floated through the air. Distant, like a constant hiss, came the sound of the fire."

A second sound could also be heard nearby: the throaty roars of Caterpillar crawler tractors. Fourteen of the powerful machines would eventually be employed at key locations on Bloedel lines, most of them hurriedly brought in by ship, truck and rail from neighbouring logging operations and from Vancouver. Given that large, efficient bulldozers were a fairly recent

*The Bloedel was the first wildfire in British Columbia to make extensive use of large, powerful crawler tractors. Note the lack of a safety canopy for the "catskinner."* BC Archives photograph NA-07023

innovation in logging, the Bloedel fire became the first major BC blaze in which they played a significant role. They cut 480 kilometres of fireguard around and through the burn, the most extensive such network ever built in this province. Men riding these bucking machines were counted among the fire's heroes. Among them was Roy Piercey, about whom a firefighter admiringly told Anderson, "That boy sure can sit a Cat." Harold Crummer, the Vancouver tin-canner, was awed by the catskinners' endurance.

> Crummer: It was hot work, and it was *so* smoky. The bull-dozers would go into that smoke in the morning and you wouldn't see them until they came out in the evening. I don't know how them guys stood that. It was just like a fog. It was terrible.

By clearing and defending a strategic line west of Headquarters, firefighters prevented flames from directly attacking the little company settlement, despite the fact that the fire activity of July 22 and 23 was the most spectacular to date. By then, approximately 1,500 firefighters were toiling on fire lines around a 150-kilometre perimeter.

Among the men in Comox Logging's employ was 21-year-old Hi Churchill. Born on a hardscrabble Nova Scotia farm, Churchill had moved to Vancouver as a child after his father died in the 1918 flu epidemic. In the Depression winter of 1935 the 18-year-old Churchill had found work as a gandy dancer (a track layer) with Comox Logging. Since then he had put in a stint setting chokers—hazardous work that killed his younger brother—but in 1938 when all Comox Logging employees found themselves fighting fire, he was working as a brakeman.

Churchill, fireman Joe Ducca and engineer Bert Woodruff made up the crew of a Baldwin Two Spot locomotive that was ordered by logging manager Bob Filberg to retrieve a steam donkey threatened by fire north of the Oyster River. They crossed the upper Oyster over a bridge that firefighters had been pumping water on for three days, made the five-kilometre trip north to the Cariboo Creek logging area, hooked onto the donkey and headed back south before an advancing fire front. Arriving at the Oyster River again, they found that airborne sparks had ignited the bridge—their only avenue of escape. They later learned that, in a replay of events at Forbes Landing, the

# — Canning the Tin Canners —

Within days of their arrival at Campbell River on July 15, 1938, the "tin canner" contingent of 211 men was being publicly maligned for its lack of commitment to the Bloedel emergency. "Green hands are mixed with old loggers," wrote Torchy Anderson in the *Vancouver Daily Province*, "and your old logger will fight fire until he drops. There are stories of men among the imported crews who have shown some inclination to discuss political theories on the fire line. There have been resultant kicks in the pants and pokes on the nose from exasperated old hands." When Forbes Landing Hotel finally succumbed to flames, the *Province* ran the headline, "Sabotage Hits Fire Line; Hundred Men Dismissed; Forbes Landing Razed." The one water pump that could have saved Forbes Landing, Anderson wrote, "was put out of business by putting sugar in the gas tank." One thousand feet of fire hose, Anderson claimed, had also been destroyed through "clear cases of sabotage."

But it remained for the province's premier to level the most public condemnation of the Vancouver recruits. In a provincial radio address, Duff Pattullo told British Columbians that the Forest Branch had "definitely established" sabotage of pumps and hoses. "Malign influences are at work in our midst," Pattullo inveighed, in an accusation as hyperbolic as it was vague, "and it behooves all law-abiding people to lend no encouragement to the activities of those who would wield such influences."

One response to sabotage charges was made by Comox Co-operative Commonwealth Federation (CCF) MLA Colin Cameron. Writing in the party newspaper *The Federationist*, he said he had "run up against a stone wall" in attempts to confirm accusations. "Time and again I have found that the man who was supposed to have seen hose being cut did not see it himself but knew of someone else who had heard of another man who had seen it done," Cameron wrote. "I am still looking for that other man." Forest Branch personnel eventually identified faulty lubrication directions as the cause of breakdowns in several pumps, the respected MLA noted. Furthermore, tin canners had never been given responsibility for fire pumps, which had been run exclusively by loggers, locally hired men and, late in the fight, by sailors from the Royal Canadian Navy destroyer *Fraser*. Cameron acknowledged that many of the Vancouver men did slack off on the job.

Cameron: So, on their own say, did many of the local men and local loggers hired by the companies. Let us grant that the chances are that the single unemployed did more and better loafing— should we be surprised? Many of them are in no fit physical condition for work of this nature. One lad was taken to the hospital suffering from cramps and the case was diagnosed as prolonged undernourishment. In addition to this, it seems hardly reasonable that men for whom society seems to have no use should unduly exert themselves for 25 cents an hour, less a dollar a day for board, in order to save the assets of corporations who have been allowed to possess themselves of a large share of our natural resources, and whose treatment of those resources has been such as to bring the major industry of this province to the verge of exhaustion.

One cannot help but feel that the province of British Columbia received from these men far more than it deserved. A campaign [of accusations] does, of course, serve a certain purpose. It tends to divert the public mind from the real causes of this disaster and the real criminals responsible. Someone is responsible for logging being carried on this last month— someone is responsible for the thousands of acres of slash—the accumulation of years of neglect, which has provided the fuel to sweep this fire down the island. It is to be hoped that the people whose responsibility this is will not be allowed to sneak off scot-free under the convenient smoke-screen of "sabotage" headlines.

*Hi Churchill worked as a brakeman on a logging train that ran a fiery gauntlet to save fellow loggers trapped by flames.*
Keith Keller photograph

all-important pump had broken down minutes after they'd crossed the river outbound. Churchill and his crewmates took their locie and donkey engine across the smouldering structure, then pulled into a nearby siding where approximately 75 anxious loggers were contemplating their future.

Churchill, now 90 years old and living near Courtenay, recalled how that afternoon unfolded. "The men didn't know what to do, so they had a vote to see whether they wanted to stay where they were at or whether they wanted to go in to Headquarters. And they decided that they would go in." But before making a move, the loggers were forced to await the arrival of a larger locomotive, a Baldwin Three Spot, which had been dispatched from Headquarters to rescue them. They had no way of confirming its whereabouts along the track because company telephone lines had already burned.

When the second locie arrived and the men began preparing for their flight southward, another problem became apparent—there wasn't enough room for everyone in the crummies, the railway cars used to transport loggers to and from their work sites. The Three Spot was pulling a water car, however, and around the car's tank ran a wooden boardwalk. Once the crummies had been filled, the remaining men squeezed onto that boardwalk. They pumped water over themselves, then maintained a protective spray as the little convoy headed down the track. As precarious as their situation was, Churchill remembered the group remaining calm as they prepared to run that fiery gauntlet.

Churchill: The Two Spot was the lightest engine, so they put us in the front, and they hooked the heaviest one on the back. They did that because the ties were burning, and if the heaviest locie went off the track—if it broke through a burning tie—the Two Spot could still be mobile. It took quite a while, because with the ties burning you just kind of crawled along. If you went rolling along too fast some of the burning ties might have broken. There was fire on both sides of the track. Once we got to an area close to Camp Three, the fire had made a sort of a curve and was going the other way, towards the hills, so we were out of the fire then. We made it to Headquarters, but the steps on the locomotive and the crummies were all burnt off when we got out.

Upon arriving at Headquarters, Churchill and his Two Spot crew were instructed to return to the Oyster River to retrieve the steam donkey, the object of their original rescue mission, which they'd left at the Oyster River siding. They rode back to the river in the Two Spot, hooked up to the donkey and headed back to Headquarters, successfully running the burning line for the third time.

While Churchill and his companions retreated from the Oyster River, a Comox Logging crew led by foreman Oswald Harmston had a scorching escape of their own as they attempted to protect a railway bridge on the Tsolum River. Cats had scraped a wide swath of bare soil around the structure, and fallers had removed snags for a greater distance still, but as Constitution Hill erupted in crowning flames, fire-generated winds created an inferno in nearby slash. Finally, when the loggers were forced to turn their hoses on each other for protection, Harmston gave the order to retreat. Boarding a locomotive-drawn flatcar, the men covered themselves with a wetted canvas and retreated through a blazing landscape. Though the flatcar caught fire several times and one firefighter's smouldering coat had to be extinguished by his companions, Anderson reported that Oswald's crew walked into the company office to remark that "it was pretty hot up there" and to ask for their next assignment.

Given that railway lines provided the sole access to most key fire areas, it isn't surprising that other railway crews experienced close calls. Anderson

*Fallers use a crosscut saw to begin the dangerous work of cutting a large snag on the Bloedel fire. Left standing, burning snags often became "candles on the devil's birthday cake," sending sparks flying across control lines to begin new fires. Note the large slabs of Douglas fir bark lying at the faller's feet. These slabs could fall without warning, injuring or killing loggers.* BC Archives photograph NA-06428

recounted for his *Province* readers how, on July 22, an ERT crew composed of engineer Andy Teak and brakeman "Waggo" Wayne ran "a sight-seeing tour through hell" near Camp Eight. The pair put their locie and attached tank car through a blazing section of track 15 times, on each pass hosing flames threatening to spread to surrounding slash and from there to a fortune in decked logs. A Forestry Branch official who witnessed the feat told Anderson it was "the most unbelievably brave thing I have ever seen." The pair were subsequently immortalized in a *Province* headline dubbing them "The Great Fire's Original Asbestos Men," while the laconic loggers commented in the style of the Harmston gang that "she was fair to middlin' hot" in there.

For all of the fire's fury, it remains unclear how Anderson justified his July 23 suggestion that the emergency qualified as an incipient national disaster. It may have had something to do with his perception of himself as a witness to history. It may also have come from the fact that he was a freshman reporter at the time, just beginning his 11-year rise to the editor's chair at the *Daily Province*.

However, there was no disputing that the Bloedel blaze dominated provincial news in July 1938. On July 21 the BC government had belatedly shut down all logging on Vancouver Island from Sayward to Victoria. Two days later Premier Pattullo took the unprecedented measure of banning anyone

*Inside a mess tent on the Bloedel fire. At its peak the fire was fought by 1,500 loggers, local residents and Vancouver recruits.* BC Archives photograph NA-06454

from entering forests in that area or on the mainland coast south of Knight Inlet. In a public radio address, he appealed to British Columbians to seek their recreation at beaches rather than in the province's tinder-dry forests. A record 110,000 people were reported playing on Vancouver beaches the following day.

In settled areas immediately east of the great fire, going to the beach had taken on a more sinister significance. With no roads running north from the town of Campbell River and with flames arcing southward around it, a marine evacuation of the town became such a real consideration that the destroyers *St. Laurent* and *Fraser* steamed north from the Esquimalt naval yard in preparation for a mass exodus. Anchored off Campbell River, the ships sent ashore teams of telegraphers who, from a triangle of bases at Campbell River, Camp Eight and Headquarters, provided invaluable communications around the extensive fire zone.

By July 23, as the risk to the Comox Valley overshadowed that at Campbell River, both destroyers shifted south to stand off Comox. A third ship, HMCS *Armentieres*, landed a large supply of water buckets at Comox before proceeding north to provide continuing coverage at Campbell River. Evacuation by sea proved unnecessary, though the fire did reach tidewater at Oyster Bay, midway between Courtenay and Campbell River. Sailors nonetheless provided assistance by patrolling Courtenay's ash-strewn streets against spot fires and by helping evacuate endangered farmers around Merville, and Mennonite settlers at Black Creek, some of whom had buried farm equipment and, in one case, a piano as insurance against the approaching flames.

*Naval telegraph operators provided vital communication links within the massive fire area. The operators were sent ashore from ships standing by in case an evacuation by sea should prove necessary.*
BC Forest Service collection

Over two and a half weeks, the beleaguering northwest

*Mennonites evacuated from their Black Creek farms hold an outdoor service in a Courtenay park where they live in a makeshift tent village.*
BC Archives photograph NA-06417

wind had driven the fire from north of Campbell River to the outskirts of Courtenay, the majority of that distance having been covered in four days. When the narrowing front entered heavily treed lands below Forbidden Plateau, the great unknown was: how far south would flames run in the prevailing wind and heat? Its scope for southward travel seemed almost unlimited, and several communities in that direction had been placed on evacuation alerts. However, firefighters and residents of the threatened region awoke on the morning of Sunday, July 24, to calm weather and a fire behaving accordingly. Though flames continued to advance, "widening the reeking path which it has ripped through the heart of the Island," in Anderson's evocative words, fire managers recognized their opportunity to deal the Bloedel blaze a significant blow. Whereas searing heat had been forcing crews to build control lines well back from flame fronts, Cats were now pushed forward to the fire's edge. Two days of toil concluded with supervisor

# — Wildfire and Wildlife —

In 1938 the Campbell River area was as renowned among hunters for its deer and blue grouse as it was among fishermen for its trophy tyee salmon. Before the Bloedel blaze had been completely extinguished, chief forester Ernest Manning ordered a preliminary report on the conflagration's impact on game birds and animals. Though the blackened landscape would soon rebound with shrubs and berry crops that in turn attracted great numbers of grouse and deer, the Great Fire had taken a substantial toll.

*Journalist Harold "Torchy" Anderson with a fawn orphaned by the Bloedel fire. Anderson's scathing reports in the* Vancouver Daily Province *helped condemn on scant evidence the contingent of unemployed "tin canners" who had been sent as an emergency firefighting force.*
BC Archives photograph NA-06427

Game warden Barney Harvey, involved in firefighting in the Campbell River area, noted that deer had sought shelter in swamps, many of which were dry, offering little protection from flames. Around Comox Logging's Camp Three, however, several large swamps remained wet and provided effective refuges.

Fire crews in the Quinsam–Elk Falls area found many dead deer as well as surviving animals suffering various degrees of burns. Forest ranger Charlie Langstroth remarked on the "strong smell" of rotting carcasses along the Forbes Landing road. Fire inspector Douglas Taylor told how, on one volatile day, he counted approximately 75 deer "milling around" in a slash-filled gulley south of Lower Campbell Lake. Fire encircled the group, "and since none came out, it is assumed that all were killed." Some animals not killed immediately fell victim to hot embers. Many surviving grouse had their toes burned off; Taylor remarked on seeing a similarly afflicted black bear at the Tsolum River bridge near Courtenay, "with forepaws burned so badly that it whimpered every time it set one down."

Firefighters consistently reported observing a peculiar behaviour among spruce grouse. Foreman Harold Carter noted, "A curious fact...was that the grouse would fly up and hover...and would then suddenly dart into the fire as if a sudden draught had caught them."

Animal losses were particularly high among the year's young, a fact that provided Torchy Anderson with a classic Torchy sidebar story. "Last night about 100 yards from where a roaring hell blocked a logging road I saw a young fawn, so young it could scarcely toddle on its spindly legs, looking for its mother," he wrote. "I tried to get around it in order to direct it away from the fire, but it persisted in moving in the wrong direction, occasionally lying down on the road through sheer exhaustion. Unable to do anything, I had to leave it. My last sight of the little fellow was his tiny figure silhouetted against the spark-belching underbrush."

Bert Conway declaring that guards across the critical southern head were "consolidated." The outstanding wild card was wind, and when over succeeding days barely a puff of breeze disturbed the smoke pall cloaking eastern Vancouver Island, firefighters knew the end was at hand. There remained a mammoth task in the form of mop-up and patrol, but the Bloedel fire, Vancouver Island's Great Fire, was dead in its 50-kilometre track.

The Great Fire's cause was officially listed as "unknown," though there were hints of foul play after it was learned that on July 5, the day the Bloedel fire ignited, two Forest Branch lookouts had noted in their logbooks that the smoke was black enough that it could have been issuing from an oil fire. Nothing, however, was proved. As for the fire's extent, though Torchy Anderson's emotion-wrought dispatch of July 23 had put its total sweep at 518 square kilometres, accurate surveys later proved that estimate to be about 40 percent too large. Still, 30,000 hectares was a lot of scorched ground, and the cost was monumental.

Comox Logging alone lost 40 million board feet of prime felled and bucked wood, twice the volume of all other companies combined. The majority of standing timber killed also stood on Comox Logging land, obliging the company's crews to spend several years salvaging what they could of the charred trees. Though this standing timber occupied only 15 percent of the fire area, the amount of merchantable wood affected totalled 460 million board feet, of which 80 percent was considered salvageable. The resulting forfeit of more than $400,000 in stumpage fees was by far the province's most significant loss, four times greater than the suppression costs borne by BC taxpayers.

As firefighters turned in their pumps and mattocks, lawyers and corporate heavyweights took up positions on fire lines of a different sort. Two major bottom-line battles emerged, each involving Bloedel, Stewart & Welch. In the first, BS&W—described by forest historian Ken Drushka as "probably the biggest, most modern, most aggressive company on the BC coast"—attempted to recover a substantial portion of its firefighting costs from the BC Forest Branch. At the same time, the company faced a lawsuit filed by the Elk River Timber Company (ERT), which demanded compensation for its suppression costs and fire-related losses. On February 20, 1940, having spent seven weeks listening to both companies press their cases, and having spent 40 hours

*A charred landscape stretching to the horizon—the immediate legacy of the Great Fire.* BC Archives photograph H-038543

deliberating, a BC Supreme Court jury announced its decision in favour of ERT. Justice Aulay Morrison subsequently awarded ERT more than $90,000 in damages. Comox Logging, despite losing hugely in standing timber, cut wood and infrastructure, did not press a similar suit against BS&W.

As the Supreme Court drama was proceeding, BS&W waged a diplomatic but at times testy battle against a Forest Branch decision that the company should bear firefighting costs on all its claims. Lawyer John Wallace deBeque Farris, King's Counsel, argued for the company that the portion of the fire situated north of Lower Campbell Lake was BS&W's responsibility but that everything south of the lake was a geographically separate and therefore unrelated fire. BS&W had spent $66,000 fighting that southerly fire, he said, and they wanted their money back.

The Forest Branch countered that the Bloedel fire simply blew across Lower Campbell Lake to ignite the Forbes Peninsula and from there continued its southward spread. BS&W claimed that the southerly fire could have been touched off by tourists. When that dubious thesis fell flat, the company argued that backfires set by ranger Langstroth on the Forbes Peninsula were responsible for the blaze's advance, a contention they did their best to support using information gleaned in gruelling pre-trial interviews with Langstroth.

However, BC chief forester C.D. Orchard—Ernest Manning's successor—pronounced the Forest Branch's final decision late in 1942 when, in a curiously pedantic letter, he lectured BS&W's management at length on wildland fire as it relates to the forest industry. When the chief forester arrived at his point, he erased any hope BS&W may have had that the province would pay the $66,000 claim. "Your fire of 1938 was distinguishable from all other industrial fires only in its size and disastrous results attributable to extreme fire weather conditions. The principles are the same regardless of cost and loss. Fire inevitably will occur on logging operations and the costs involved in firefighting, fire damages, loss of felled and bucked timber, and in fire prevention measures must be deemed an operating expense which every operator must be prepared to pay."

# CHAPTER 3

# The Battle of Midday

O n July 18, 1960, the day the Dean fire doubled in size to 3,200 hectares and achieved headline status, Harold Larson and his brothers were baling hay on the Pine Ranch, along the Coldwater River south of Merritt. Making hay wasn't a problem that summer—the sun shone relentlessly, curing hay and creating temporary employment for upwards of 4,000 firefighters toiling throughout the province. The problem, from the Larsons' point of view, was getting the time to accomplish everything that needed doing in the course of managing their overly busy lives. Sunup to sundown never seemed to be quite enough. Asked decades later whether they had any photographs of themselves from the era, they looked up as though they had been asked whether they'd spent much of the period in the casinos of Monte Carlo. "No photographs," Harold said flatly. "We had no time for photographs."

# The Battle of Midday (The Dean Fire, 1960)

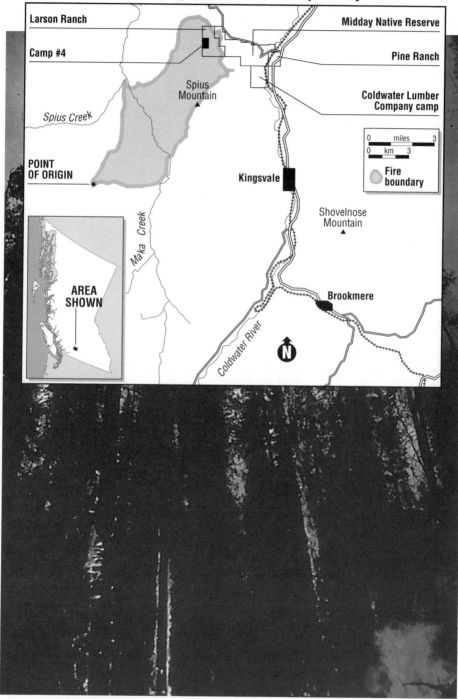

*The ranching Larsons photographed at the Rey Creek Ranch near Logan Lake, which Harold Larson now manages. From left to right: Violet, Wendell, Rosemary, Harold.* Keith Keller photograph

The Larson brothers, Richard, Harold and Wendell, ran their own 300-hectare place in the Midday Valley south of Merritt. They had moved there in the late 1940s, 10 years after leaving their Alberta homestead, and they had been working like hell ever since to make a go of it. Harold, his wife Violet and their three children lived on the Midday Valley place, along a dirt road that turned away from the Coldwater River at the Pine Ranch. Along with Richard, the eldest brother, Wendell and wife Rosemary—Violet's sister—lived at the Pine, which belonged to Chris Christopherson of Seattle, as did a couple of other parcels in the Midday; the Larsons ran the Pine while simultaneously building up their own place. Larson cattle and Christopherson cattle ranged together into the Coquihalla country as far south as a height of land now known as Larson Hill.

Until July 18 the Larsons had been aware of the Dean fire, but it hadn't become a central element in their lives. That was about to change.

In that summer of 1960, wildfire had burned itself into the consciousness of people throughout western North America. States of emergency had been declared from California to Washington, and the BC Forest Service was stretched to the limit in dealing with hundreds of blazes, the majority of them burning in the sprawling forest region centred in Kamloops.

The people of Merritt, among them the district's forest ranger, certainly had fire on their minds that summer. Don DeWitt remembers 1960 as "the year I got the tail burned off me." This was at first due to the Nic, which blackened a vast tract of valuable range land above Nicola Lake. (Tradition has it that forest fires are named by taking the first three or four letters of a feature, usually geographical, associated with their origin.) The Nic was one reason that folks around Merritt were, fairly or not, developing a dim view of

*Retired forest ranger Don DeWitt, left, with former assistant ranger Don Dearing. DeWitt recalled 1960 as "the year I got the tail burned off me," on the Dean and several other major fires in the Merritt area.* Keith Keller photograph

the BC Forest Service. Ranchers were unhappy that the fire had happened in the first place. They were even more steamed when the Forest Service torched a large swath of grass ahead of the fire to prevent its advance. DeWitt found himself at the controversy's epicentre. "It seemed to me that when I left Merritt everybody was mad at the Forest Service. We put 90 percent of the fires out or contained them, but the Nic burned a lot of winter range for the farmers. It was grazing land all around it, and that whole mountain burned."

Then, overlapping the Nic and several other challenging fires, came the Dean, which began 25 kilometres southwest of Merritt on a branch of Spius Creek, which flows out of rugged hill country between Boston Bar and the Coquihalla Pass. Initial reports referred to it as the Spius Creek fire, so its Forest Service moniker could easily have been the Spi. It could also have been called the Nad, as the Nadina Lumber Company was one of many small logging and sawmilling outfits cutting lumber in that country; sparks from the open burner the company operated to dispose of wood wastes touched off the Dean fire. But the fire's enduring name evolved by accident, so the story goes, when a garbled radio conversation resulted in a Forest Service dispatcher jotting down an unusual contraction of Nadina.

"We should have known that it was getting dry," DeWitt said, reflecting on the fire's origin, and the fact that—as with the Bloedel disaster—an industrial closure may have prevented the Dean. "That hot wind blowing all the time dried things out faster than I thought it would. Anyway, one day she took off, and away she went." Nadina crews attempted to control the outbreak themselves until it became obvious that the rapidly spreading fire demanded further resources. Loggers from other operations were rushed to the scene, and for a week they attempted, and failed, to control the Dean.

> DeWitt: The country there is broken up into draws, and it was the wind that was the killer. The prevailing wind is from the south. But in the side draws, in the valleys on either side of the Coquihalla for 20 miles, the winds could be from the south to the north or the north to the south or any other way. Local winds.

It was these local winds that sent the Dean raging out of the Spius Creek drainage and over Spius Mountain on July 18. As flames bore down on the

Midday Valley, truck driver Floyd Brown stopped his rig beside the Pine Ranch hayfield and told the Larsons they should consider evacuating Harold's family. The brothers dropped their haying, jumped in their '52 Chevy pickup and sped off—to whatever extent it was possible to speed along the narrow road snaking into the Midday. Arriving at the farmhouse, Harold found his youngest child asleep in his crib and no one else in sight. His first response, he recalls, was something along the lines of "Aaagh!" He tracked the remainder of his family to a nearby knoll where Violet had gone to assess her predicament. Alone with her children and with no means of escape, she was on the verge of panic. The roar of the fire was clearly audible. Police arrived and, though the family needed no prompting, ordered them to evacuate to the Coldwater River, which they did after throwing a few essentials into their truck.

*Volunteers sign up for firefighting work on the Dean fire. The intensity of the 1960 fire season eventually resulted in a shortage of firefighters that could only be resolved by forced recruitment.* BC Archives photograph NA-20166

The BC Forest Service responded to the Dean's wind-driven outburst by rushing additional firefighters and 10 bulldozers into the Midday Valley. On ground rented from the Larsons, a tent village known as Camp Four was thrown together—makeshift accommodation for 500 hastily recruited fire-fighters. The Midday Valley was strategically critical, being home to the Midday Valley Native reserve and the last point of road access for firefighters between the Dean's five-kilometre front and the town of Merritt.

As the fire added to its perimeter, a Forest Service official stopped by the Pine Ranch, hired Wendell Larson as a section foreman and sent him to the front lines. Richard stayed behind to manage things on the Pine. Wendell had charge of a crew of conscripts bussed in from the Okanagan. He wasn't overly impressed with his work force. "The only time you'd see them," he quipped, "is when the lunch truck drove up." His skepticism was fed by rumours that people had seen shoes being thrown from the bus window as the recruits were being driven to the Dean fire camp, a ploy used, he presumed, to delay or prevent their being sent to the fire lines.

Though Wendell Larson was put on the Forest Service payroll, his brother Harold had been warned that as an occupant of the Midday Valley ranch he would be required to fight the Dean as an unpaid volunteer should the fire enter onto Larson land. It did, and Harold rode into battle aboard the family bulldozer, an International TD-9 that he describes, perhaps over-modestly, as "a glorified farm tractor."

The battle of Midday was three days old when Harold's bulldozer blew a seal, forcing him to drive to Vernon for parts. He was on his way home, driving his Volkswagen Beetle across Douglas Lake Cattle Company land around midnight on Wednesday, July 20, when he heard a radio bulletin that was both a pleasure and a surprise. The announcement has remained fresh in his memory. "I heard the news commentator on the radio: 'The Dean fire is under control. I repeat, the Dean fire is under control.' I thought, Great!"

An experienced firefighter, he also knew that there would still be mop-up work to be done. Upon returning home, he reassembled the TD-9 and returned to the fire. He and his brother were working together when all semblance of control suddenly evaporated. He was still on the Cat while Wendell continued to supervise "that bunch from Kelowna," the Okanagan recruits he'd come to hold in such low esteem. When flames got up into the trees and

*Bulldozer operator Joe Horvath had a close brush with death when the Dean blew out of control near Camp Four. Unable to escape to safety, he hid under his Cat until the blow-up passed.* Photograph courtesy Joe Horvath

began crowning over the men's heads, Harold and Wendell decided it might be a good idea to look for a way out.

Joe Horvath was thinking similar thoughts. In 1960 the Austrian-born Horvath was 24 and running a D6 bulldozer for the Coldwater Lumber Company, whose bush mill was located across Spius Creek from the Nadina Lumber Company show. Coldwater had early on sent men and machinery to the fight, among them Horvath and his Cat, and when the Dean blew he was cutting guard a short distance from Harold and Wendell. The escape route he chose was one he would quickly come to regret. In the late 1950s the Larsons had sold the standing timber on their Midday Valley place rather than continuing to cut and mill it themselves while simultaneously running their own ranch and the Pine's diverse holdings. Nicola Valley Sawmills had finished logging this timber in the winter of 1959, leaving behind, in the form of slash, a fabulous fuel supply that the Dean was now consuming. Disoriented by thick smoke and flames, Horvath drove his Cat along a road running through this cured slash. By the time he realized his error, flames had cut off any avenue of escape.

> Horvath: I thought the end had come. The heat was terrible. There was fire on both sides of me and over my head. I thought I would make it to the camp but the heat and smoke just got so I couldn't breathe any more. The fire was roaring, and it was blowing dust and everything in the air. You couldn't see where you were going. I knew there was a meadow around there, Larsons' Meadow. I wasn't very far from it, but I didn't know that. I finally lay down under the Cat with my face down close to the ground and breathed the oxygen there. I was by myself for a long time. I could have got back on the Cat and tried again, but there were a lot of trees down and they were burning.

John Dearing, an assistant forest ranger working under Don DeWitt, was one of the first people to encounter Horvath after he emerged from under his Cat. If anything indicated the heat and smoke that Horvath had encountered, Dearing said, it was the condition of his eyes. "I've never seen anyone's eyes look like that before. They looked like two tomatoes. It was frightening." Dearing flicked cold water into Horvath's eyes—the only treatment available at the moment—then helped him into a vehicle heading toward Merritt.

The firestorm that struck around Camp Four caught Dearing and a large contingent of firefighters by surprise. Its intensity was overwhelming.

> Dearing: The fire came with a sort of vortex—cyclonic winds. The afternoon had been sunny and clear, and all of a sudden it turned black. Tongues of flame were shooting out from the hillside, whipping past us. Men went flying. I'll never forget those guys flying up in the air, maybe 20 feet. And man, they hit hard when they came down. Some of the guys got hurt—not badly, but sore ribs and stuff. I didn't think that oil drums would move, but they went up too. And boards—pieces of wood became lethal objects. It was

*Camp Four, built on the Larson ranch to house 500 firefighters, was destroyed by a spectacular blow-up on the Dean blaze.* BC Archives photograph NA-20217

pandemonium, with all the noise and the dust and the stuff flying around. Things were bursting into flame, men were running everywhere. Some guys were tearing their clothes off as they ran; people will do weird things under those conditions. The whole camp burned. I went back the next day to see if I could find any of my own stuff. I found a bit of my sleeping bag zipper. Everything else was fricasseed.

As the blow-up advanced, heavy ground smoke prevented Wendell Larson from fully appraising the plight that he and the men in his charge were in. Sensing that his crew would be at least temporarily safe on the fireguard, he instructed them to stay put while he and Harold rode their bulldozer to Camp Four to size up the situation. They chose a different route from the one they had seen Horvath take, correctly assuming that his path through the burning slash would be impassable.

*Firefighters back off as flames approach a bulldozed fireguard on the Dean fire.*
BC Archives photograph NA-20221

Wendell Larson: When we got to Camp Four, which was on our place, it was on fire. Nobody was there, the thing was just on fire. And our truck was gone—we had a vehicle parked there. The food truck was gone and the grub was laying there. And the tents were on fire. So we just kept on going.

Whipped by the wind, burning pieces of canvas from the disintegrating Camp Four tents flew at the bulldozer like a flock of crazed, incendiary birds.

Harold Larson: We decided to get out of there because it was getting so hot. You could see big old stumps below us were blowing up—catching fire. Wendell was soaking my cap with his water bottle while I drove the Cat. I put the cap up against my face so I could see. When that one got too hot I'd give that one to him and he'd give me his.

Even from the relative safety of the Coldwater River, where the Horvath family was living in Coldwater Lumber Company housing, the steadily growing smoke cloud and rain of ash indicated that some ominous force was being unleashed in the Midday Valley. Joe Horvath's wife Margot had gone into Merritt to establish temporary living quarters for the family and was on her way back for another load of possessions. Getting to the now-evacuated Coldwater lumber camp required her to drive through the Pine Ranch. That was as far as she got.

> Margot Horvath: There was a roadblock there—police or Forestry, I forget which. They told me, "No one can go through. The fire has jumped the guard and some of the men are trapped in there."
>
> I said, "My husband is up there!"
>
> They said they were bringing some of them out. I forget whether it was a car or a pickup, but I could see Joe sort of hanging in the middle of some other men. His face was black. All I could see was his eyes and his teeth. They were going to take him to hospital. I said, "I'll do that myself." So they transferred him into my vehicle and I took him to the hospital.

Joe Horvath spent one night in hospital recovering from smoke inhalation before returning to the Dean fire.

Violet and Rosemary Larson were also waiting at the Pine Ranch for news of their husbands. What they saw next was, in its own way, as shocking as the scene that had greeted Margot Horvath. A convoy of vehicles poured out of the Midday Valley. Among them were Harold's Beetle and the family truck. There wasn't a Larson at the wheel of either one of them.

> Rosemary Larson: Floyd Brown came to our place when all the vehicles were going past the house where Violet and I were. He phoned the police and told them to stop the vehicles before they got into town because they were running out on the fire. They had our truck and Harold's Volkswagen. Violet and I could see all the vehicles going by—and *our* vehicles going by—but nobody told us what was happening.

*Margot Horvath on the doorstep of the family's company housing near Merritt in 1960. She drove husband Joe to hospital after he escaped the Dean fire. "His face was black. All I could see was his eyes and his teeth."*
Photograph courtesy Joe Horvath

Left in the Midday Valley without a vehicle, the Larson brothers found themselves among the many firefighters considering their next move. Some were running flat-out away from the fire area toward what they hoped would be safety. Having determined the fate of Camp Four, the brothers began walking back to check on the men they'd left on the fireguard. Partway there they encountered Floyd Brown's lumber truck rumbling through the smoke. Brown, a one-man evacuation force, had loaded up a group of Kelowna recruits and driven them out of danger. (A go-getter, Brown would go on to become mayor of Merritt.) The Larsons suggested that he go back for Wendell's crew. Brown replied that he had another mission on his mind, and he drove away. "He was busy," said Wendell. "He was hauling men out and he was trying to put the fire out. Same with all the loggers and the local people. They was trying to put it out. It was the guys that come to the fire barefoot that were running helter-skelter down the road."

*Marking hard hats with BC Forest Service identification on the Dean fire.*
BC Archives photograph NA-20172

A forest officer drove up at that point, and the Larsons appealed to him to retrieve Wendell's crew. "He drove back there a little ways, then he came back with tears in his eyes," Harold recalled. "He said, 'I'm afraid I can't get to those men.' And he took off back the other way." The Larsons continued walking in, meeting, partway there, the men they'd left behind making their way out of the fire.

As often occurs, the Dean blow-up was triggered by a shift in weather systems. As that shift subsided, the fire's intensity calmed, its area by this point being 4,900 hectares. The Forest Service established a new camp in an alfalfa field that the Larsons had rented from the Midday Valley Native reserve. It had been to that field that John Dearing had directed fleeing firefighters as they ran in panic from the firestorm that struck Camp Four. Getting a meal into the panicked men helped calm them down, he noted. "From then on it was just the dull drudgery of re-establishing everything. And smoke and tiredness and dealing with people." Among the people Dearing dealt with that night were people whose health had already been tenuous when they were sent to the Dean, and whose conditions were aggravated by the shock of their narrow escape.

> Dearing: That night we had six or seven people with seizures—borderline diabetics. The stress overcame them, which isn't surprising 'cause some of them had thought they were going to die when the fire and that wind hit us. That night in camp, someone would say, "There's something going on in the grass over there!" You'd go over with a flashlight and there'd be a guy thrashing around. Fortunately, we treated them properly, and we got them to the hospital. No one died.

Though no lives were lost on the Dean directly, one death occurred on the fire's periphery. He may have been someone looking for a job, though no one knows for sure. He showed up at the Pine Ranch one day asking directions to the Dean fire camp. After getting an answer from the Larsons, he ended up on the Canadian Pacific Railway tracks along the Coldwater River and was run over. The train crew stopped, loaded the corpse into the baggage car and hauled it into Merritt. That was the last the Larsons heard of him; by now they had troubles of their own to consider.

Though the majority of their Midday Valley place had been burned over in the Dean blow-up, some turn of fate had left their house untouched. When Wendell and Harold returned to it after the blow-up, they found a firefighter sitting out front, playing the guitar Harold had left behind when he'd evacuated his family. Harold put a quick end to the strumming.

An old barn, weather-cured and full of dry hay, had somehow avoided the storm of wind-lofted embers. A new log barn under construction, however, despite being nothing more than a roofless and floorless shell, was incinerated, as was the chicken house and a storage shed.

> Harold Larson: We had a little shed full of old power saws, kegs of nails and stuff. We had thought that the fire might come through, so we'd set the stuff out on the dirt road.

Somewhere between when we moved out and when the fire jumped, somebody drove down the road with a Cat and drove over this stuff that we were trying to save. We thought that this was no good, and the fire was finished now anyway—under control—so we put it all back in the shed. Then the shed burnt. So instead of just having bent nails, we had burnt, bent nails.

Aggravating their situation was the discovery that, as Harold later described it, "What the fire didn't do, the firefighters took care of." The family had been on the verge of harvesting their Midday Valley alfalfa crop when the Dean interrupted their ranch work. Re-establishing Camp Four in that field ruined much of the crop. That wasn't the end of the troubles.

Rather than recover a cache of diesel fuel left over when the Dean was out, someone in the Forest Service apparently decided it would be simpler to punch holes in the drums, drain them and bury them on the Larson property. It takes a long time for anything to grow back on diesel-saturated ground, Wendell noted. The brothers' woes didn't end there.

Harold Larson: I guess some firefighters got bored 'cause they practised throwing their axes at our fenceposts. I'll bet there wasn't a space for two miles where there was anything left on the post. We'd put in humungous big fence posts to make the fence last. And every post had the barbed wire cut off alongside the staples.

The Forest Service eventually paid Wendell wages for his time on the fire, though his time slips were lost and he had to sign an affidavit before that was settled up two or three months later. They also paid for the lost alfalfa crop, but balked at compensating Harold for the time his TD-9 worked. A compromise was finally worked out on that one.

Harold Larson: They didn't want to pay us for our Cat. The maddening thing was that there was two D8s and a D9 from one construction company that these great big lantern-jawed geezers was driving. And they marked down

*Obscured by smoke, the* Marianas Mars *unleashes its payload of water on a Dean hot spot.* BC Archives photograph NA-20179

18 hours a day. There's no *way* they worked 18 hours a day. But when the timekeeper tried to get lippy with them, this one guy grabbed him across the counter and said, "What time do you get up in the morning?"

The timekeeper said, "Well, I'm on the job at eight o'clock."

The Cat guy said, "Then you ain't up early enough to see what time we go out. It's 18 hours we worked." The Forestry didn't have no trouble paying *them*.

Wendell Larson bought his brothers out the year after the Dean fire but sold the ranch shortly thereafter. The day that Violet and the three children were evacuated was the last day that a Larson lived on the Midday Valley place. "As far as I was concerned, it wasn't worth it," Harold commented. "There was all the damage the firefighters did, plus we lost all our grass. After that fire there was nothing left but bare earth."

*Pilot Alex Davidson does a pre-flight check in the cabin of a Martin Mars water bomber. In 1960 Davidson and aviation pioneer Dan McIvor used the* Marianas Mars *to help fight the Dean fire. The flying boats had just been converted for firefighting work, and the event marked the first time that one of the innovative Vancouver Island-based aircraft saw service on the BC mainland.*
BC Archives photograph I-51529

# CHAPTER 4

# The Flood From Above

In early July 1960, one day after the Dean fire ignited in Spius Creek, a dry lightning storm careened through central BC, igniting hundreds of new wildfires. Soon between 3,000 and 4,000 firefighters were sweating it out on fire lines in all corners of the Kamloops forest region. They were supported in the air by an innovation that had only recently become available to the BC Forest Service: fixed-wing tanker aircraft.

BC's air tanker program, now considered one of the world's best, was in its infancy in 1960. Art Seller of Langley Air Services played a key role in getting the thing off the ground when he purchased several Stearman crop-dusting biplanes. Tiny and manoeuvrable, the Boeing-made Stearman was also helpful in a limited way in forest-fire situations. According to retired BC Forest Service employee John Weinard, who helped pioneer the province's adoption of aviation as a firefighting tool, "If you could get two or three [Stearmans] working together on a smallish fire, they had some reasonable effect."

The problem was that not all fires were smallish, certainly not in the trial by fire that 1960 became. And even with small fires, 680 litres of water didn't pack much dousing clout. Linc Alexander—his name was Al Lincewich in those days—flew Stearmans in the early stage of his long aerial firefighting career. A former Royal Canadian Air Force pilot, he was "an extremely intelligent person," by Weinard's measure, "a real go-getter. He wrote a book about air tankers, a very well-written book and an excellent treatise on tactics." Alexander, now retired from flying, is less sanguine than Weinard about the biplane's effectiveness.

> Alexander: To discharge 150 gallons of water in three seconds is not concentration enough to even reach the ground in a 150-foot canopy. If you're 100 feet above the canopy—which places you about 250 feet above the ground—what reached the ground was practically zero. So it was totally ineffective. Besides, the fires would sometimes be 20,000, 30,000, 40,000 acres—these walls of fire rolling along. They'd point you at that monstrous column of smoke and say, "See what you can do with it." That was their term: "See what you can do with it." You can't blame the Forest Service. Nobody knew any better.

Actually, Dan McIvor thought he knew better. At any rate, he had an idea. McIvor had extensive bush-flying experience, including five years with the BC Forest Service, but by 1958 he was MacMillan Bloedel's chief executive pilot. It was a scorching year in BC's forests, with more area burned than ever before in recorded history and more burned than in any season since. What's more, 1958 was the worst in a series of unusually bad summers. Mac Blo convened a group of industry reps that fall and asked what might be done in the name of collective self-defence. The problem, they agreed, was the inadequacy of aircraft engaged in fire-bombing. McIvor, who had been considering this problem for some time, presented his conclusion: only large flying boats or amphibious aircraft could provide the bombing clout they needed.

The heyday of flying boats had already passed, however, and most of the suitable aircraft either no longer existed or at the very least were in no condition to fly. Then McIvor learned through a chance comment that the

American navy was selling off its California-based fleet of Martin Mars air-craft. These were not just any large flying boats, they were the world's largest. Designed by aviation pioneer Glenn Martin, the planes had been intended for service as long-range patrol bombers but had been relegated to the exclu-sive use of the US navy as general transport aircraft.

McIvor's initial contact with the American navy confirmed that of the six Mars originally built, four of the giant aircraft still existed. Moreover, they were in mint condition. And they had just been sold for scrap metal for a total price of $23,650.

What followed, as McIvor and his co-conspirators threw themselves into saving the planes from oblivion, has become part of aviation legend. As a result of his dogged persistence, by September 1959 the world's remaining fleet of Martin Mars—*Marianas Mars, Caroline Mars, Philipine Mars* and *Hawaii Mars*—were bobbing next to each other in Patricia Bay, outside Victoria. A company, Forest Industries Flying Tankers Limited, was formed to oversee the conversion of the aircraft for their new role. That company's chief pilot was Dan McIvor.

*Marianas Mars* was the first of the fleet to be converted for water bombing, the core of her modification being a 27,000-litre tank. Filling the tank required pilots to land on water at 74 knots, edge back to 72 knots, then lower a pair of retractable scoops capable of sluicing in 30 tons of water in less than half a minute. Released at 100 knots from just above treetop level, the aerial flood could saturate up to 1.6 hectares of terrain.

Though the Mars program was developed to fight fires on lands belong-ing to member companies, the program got its first real test in July 1960 when the beleaguered BC Forest Service contacted Flying Tankers to ask for help quelling the disastrous situation around Kamloops. Two blazes were giving particular grief, the Dean and the Mil—the latter located at Mildred Lake, near Lac Le Jeune.

Money became an immediate issue. Flying Tankers told the Forest Service they could have the Mars at $2,875 per flying hour. The Forest Service swal-lowed hard and came back with a counter-offer of $1,875 per hour less. A deal was struck: the Mars would work Interior fires for one day. If the Forest Service liked what it saw, it would continue to engage the Mars, paying the full rate for the first day and any subsequent ones. If not, the Mars crew would at least receive some valuable hands-on fire experience.

For the experimental trip, Dan McIvor was joined in the cavernous Mars cockpit by Alex Davidson, chief test pilot for Fairey Aviation, the company contracted to perform the Mars conversions. Behind the pilots, responsible for a vast battery of controls, sat flight engineers Nils Christensen and Bob Morin as well as crewman George Grover.

Having made two drops on the Mil, McIvor and Davidson set course for the Dean, where ground forces were meeting with stubborn resistance from a line of fire running across the rugged terrain below Spius Mountain. "Although this area was awkward to strike from the air, it was obvious that this particular fire front had to be stopped," the pair wrote in their report. Refilling from nearby Nicola Lake, McIvor and Davidson released a series of drops on the advancing flames.

Over their radio, the pilots received immediate feedback from their experiment. "Excellent," "Bang on," "You knocked it down," came the reports. The drops allowed firefighters to complete their guards and effectively rein in the Dean along a troublesome front. Forest Service officials then had to decide whether to keep the Mars on for a second day—and incur the associated costs—or to decline its continued use. No one, it seems, wanted to take responsibility for making the decision. A gasp of great relief can almost be heard in the Forest Service inter-office memo that described how, that evening, the manager of Flying Tanker Industries phoned to report that a fire was threatening a member company's timber holdings on northern Vancouver Island. The Forest Service released the bomber from further responsibilities, assuming that any financial obligation had been avoided. A gasp of another sort emanated from ministry offices when Flying Tankers sent the province a bill for nearly $18,000 for their work on the Mil and Dean fires.

Dan McIvor left Flying Tankers in 1963 to take a management position with Pacific Western Airways. Alex Davidson went on to form Flying Firemen, a private aerial firefighting company. He was killed in 1967 while dropping water on a fire at Langford, north of Victoria. The plane McIvor and Davidson flew on their experimental 1960 mission, *Marianas Mars*, crashed in 1961 while fighting a fire near Nanaimo, killing all aboard. Another Mars, the *Caroline*, was damaged beyond repair when Typhoon Frieda clipped southwestern BC in 1962. Operating from their base at Sproat Lake, the two remaining aircraft continue to serve Flying Tankers. The BC Forest Service still calls on them for help once in a while. Today's base rate for that privilege: $11,800 per hour.

# CHAPTER 5

# By the Skin of Their Butts

By four o'clock on the afternoon of August 31, 1967, the mercury at the Lumby ranger station had risen to within one notch of 38 degrees Celsius. The relative humidity registered 17 percent. Despite the extreme forest-fire hazard, the working woods remained open for business.

Lumby's Forest Service personnel and bush crews had already been taxed by a series of lightning strikes in the drainage of Cottonwood Creek (also known by its native name, Tsuius Creek). Most troublesome of the resulting fires was Cotton #1, a blaze that made an aggressive 1.6 kilometre-long run before being controlled. Mop-up crews were still busy on Cotton #1 three weeks later when the Hound ignited. Former Lumby forest ranger Vern Hopkins described the resulting blaze as "the only fire I ever had that blew up on me and I couldn't do anything with."

# By the Skin of Their Butts (Hound Fire, 1967)

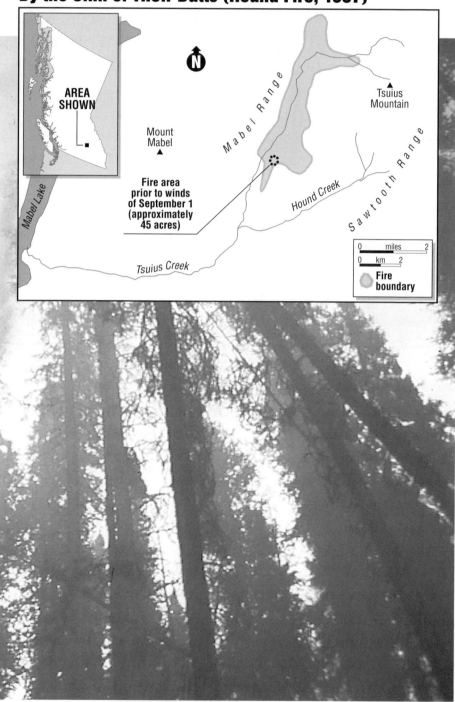

AREA SHOWN

N

Mount Mabel ▲

Mabel Range

Tsuius Mountain ▲

Mabel Lake

Fire area prior to winds of September 1 (approximately 45 acres)

Hound Creek

Sawtooth Range

Tsuius Creek

0 — miles — 2
0 — km — 2

Fire boundary

Hopkins: There had been a lightning storm in Cottonwood Creek that ignited four fires. There were two large logging operations in the area, each with 20 to 25 men to the camp. I avoided using loggers and sawmill people on fires except for on Day One. They're the closest—you get the fire, I'll get a crew from town. If it wasn't controlled, the loggers would be kept another day. Almost all the fires were controlled. Loggers were superb firefighters. They get it out and go home; there's no bullshit with them.

So we had these four fires. A couple of them were in slash. We handled them all right and had just finished cleaning them up. We'd started to bring the equipment home, clean it up and so on, and the loggers went back to work. They had 40 acres of felled hemlock, so there was a tremendous amount of fuel. And it was hot and dry. God, it was hot and dry. The fallers were working a longer shift than the others to get something done, and one faller's chainsaw started a fire. He tried to handle it himself but couldn't, so he gave up and drove to the camp for help, and then everybody turned out. Then a second crew was called to it. This was all done by the loggers organizing themselves.

The Hound—named for a tributary of Cottonwood Creek—ignited on the last day of August in an area being harvested by Ohashi Brothers Logging. After calling in a report to the Lumby ranger station, Ohashi Brothers sent runners to the nearby Cottonwood Creek Logging Company camp and to Forest Service crews mopping up Cotton #1. Within minutes, 17 men, two Cats and five wheeled log skidders were tackling the 0.1-hectare fire. A half hour after ignition, when a Cottonwood Creek crew of nine arrived on the scene, flames covered 1.2 hectares.

Loggers spent that night bulldozing fireguards and fighting the fire with stationary pumps and mobile pumps mounted on skidders. By the morning of September 1 the fire had increased to 18 hectares but was contained within fireguards on three sides and on the fourth side by the logging road providing access to the fire area. The situation was controlled to the point where, according to Monty Morris, an Ohashi Brothers logging-truck driver assigned to bulldoze fireguards, "we were just puttering around."

Morris: My swamper and I were just driving around, checking what was going on. We had no radio. We were going to come out, go down to the camp and have something to eat. We'd been up all night. This would have been late in the morning of the second day.

Morris and his fellow firefighters may have felt less complacent about their situation if they had known about ranger Hopkins' phone message.

Hopkins: About half past five in the morning I got a phone call from the duty officer in Kamloops. He had just got a message from the Vancouver weather office. He said, "You have gale-force winds forecast for your area at 12:30 this afternoon. Can you hold that fire?"

I said, "We'll go for it with all we've got."

We pushed the panic button. The two sawmills in Lumby each had 75 men on the day shift. At seven o'clock, when they went to work, they were met by school buses. So 150 men headed for this fire to beef up everything. The equipment was organized and a camp was set up because the men would be staying overnight.

The wind came up an hour after predicted: one-thirty. We had the fire completely surrounded. We had water all around it—two big pumps in the creek—so we had soaked down everything we could reach. And we had all the loggers' big equipment there. The wind blew for a couple of hours, and it was such a strong wind that the roads were completely plugged with windfall. Way in the middle of the felled timber was an area that hadn't been reached with water, and all of a sudden it flared. And away it went.

On September 1, 1967, Lloyd Siver began ferrying supplies to the Hound fire camp. The Royal Canadian Air Force (RCAF) had taught Siver how to fly fixed-wing aircraft, but in 1959 he had used his severance pay to sign up for pilot training with Okanagan Helicopters. (Among other assignments, he went on to fly choppers in minus 60-degree temperatures while working on the DEW Line in the Canadian Arctic.)

Siver: We flew in around noon. They had a fire camp of sorts and a bunch of people up the mountain. We had to go back to Lumby to pick some things up—fire pumps, hoses, whatever. I returned to the fire camp and was having coffee, and while we were sitting there the fire took off. The wind was blowing and the fire went roaring up this north-south valley. We could hear it on the radio, guys saying, "Watch out, it's movin'!"

Morris: The fire was burning above the logging road we were on, and a dry snag that was burning fell across the road. We started to angle the blade 'cause I had the blade of the Cat turned the wrong way to put that snag out. And then all at once away she went. That's all there was to it. We never even got the blade changed.

When the fire took off it started making its own wind. There was stuff flying every-where. There was lots of noise—the roar of the fire. With all that noise you didn't know what was going on around you. I couldn't go uphill 'cause it was burning back there. I couldn't go the way I came. I thought maybe I could go down the hill, but I tried and I couldn't—there was too much fallen stuff, too much windfall and things. I couldn't go up, couldn't go down, so I had to take off. My swamper had already taken off on his own.

*Bulldozer operator Monty Morris. "The fire was on both sides of me, and behind me. The only thing I was hoping was that it wouldn't come up in front of me."* Keith Keller photograph

The fire was on both sides of me and behind me. The only thing I was hoping was that it wouldn't come up in front of me. I knew there was a log landing down below, Lunzmann's landing. I thought I could get down to that. I walked down with the fire on both sides of me, about 50 feet on either side of me. There was lots of smoke. I walked toward the creek down below, probably for a half mile or so, straight down the hill. I didn't know where I was going 'cause it was all black smoke up above. When I got to the landing, it was on fire. So was the fuel tank there. It was pretty scary. I'm telling you, it was a terrible fire.

Garry Hartley was a 25-year-old faller working at Noisey Creek, north of Mabel Lake, when the Forest Service summoned his crew to the Hound fire. Hartley had cut a pilot trail for a bulldozer to the top of one ridge, then had begun falling snags on another section of the fire perimeter when, in his words, "the wind came up and brought 'er across the hillside."

*Retired faller Garry Hartley. "I never forget that fire. In a few seconds you could be dead if you don't do things the right way..."*
Keith Keller photograph

Hartley: We noticed the wind picking up and the smoke coming at us. I guess there were around 30 of us working together. We all dropped our saws. I threw my chain saw on the skidder but the skidders burned up too. There was eight or ten men that run with the fire, down into the creek. I could tell by the way the smoke was blowing that the wind was pushing the fire down into the creek. I was sure that those guys were goners. As it turned out, they made the creek before the fire did.

Siver: Those guys were running downhill and the fire was passing them, it was moving so fast. We had guys coming into the fire camp just terrified because the fire was right at their heels. They're lucky they didn't lose anybody. Anybody that would have been up that valley would have been fried for sure.

Hartley: I took off and a whole bunch of guys followed me. I'm not saying I'm no number one person, I just had a kind of instinct about the thing. You're looking after your own neck. This is getting close—you feel the heat on your face. You don't have time to think what you're going to do, you just do what your mind says at that time. I just decided that I'm going uphill. It just seemed that I took off and everybody was on my tail. Maybe they had the same instinct.

We had to go up around the fire and then walk down through the smoke. There was no place to get water to soak our shirts to put over our faces. We were maybe 30 or 40 minutes on a good run, and the rest of the time at a good heavy trot. We were exhausted. We could feel the heat of the fire most of the time until we got high enough that the fire bypassed us. But we had to come back down through the smoke to get out.

By the time we hit the trail everybody was just ready to drop. The smoke just overtakes you. That was actually the hardest part of it all—the smoke intake. In those days we were all young, so our lungs were in good shape. I wouldn't want to be the age I am now and inhale that kind of smoke. We made it out, but by the skin of our butts.

Hopkins: It was a miracle that nobody was burned. We thought 12 men *were* lost. They were on the upper edge of the fire and they could see what was happening down below. There was a high peak, Tsuius Mountain, behind this fire, and it caught this wind and caused a vortex. The fire actually blew downhill to start, though it blew every

*The Hound fire the afternoon it blew wild. It was through this cloud of dense smoke that logger Garry Hartley led a group of firefighters to safety.*
Lloyd Siver photograph

direction later on. But when it started downhill the men went uphill, above treeline, and sat it out for the night, watching the fire below them. The following day they walked way around, miles down another creek valley, and came out.

When I initially got word that we might have lost these men, I immediately put out a freeze order, not to discuss it at all, on the Forest Service radio. Can you imagine the panic that there would have been in Lumby if that kind of news had gotten out? And by gosh, nobody broadcast that news. Had any of the media picked that up, it would have caused no end of trouble for us.

Siver: As soon as we heard that the fire was blowing up, the fire boss and I jumped in the helicopter and took off. That fire had taken off and was boiling down the valley. It went seven and a half miles in about three hours. It burned from the bottom to the top on both sides, and so hot it sterilized the soil. If you flew by there a year or two later it wasn't even green.

It was fantastic, the way it boiled out of that valley. We couldn't get close to the fire because of the winds. We flew along the ridge on the back side. We were flying down ahead of the fire and off to the side, looking for spotting, because it must have been throwing sparks for miles.

Hopkins: I flew to the fire by helicopter the evening that it blew. A tremendous convection column was going up. The wind caused the column to lean, and it was leaning strongly. We flew over the fire, which meant we flew under the lean of the column. These columns pick up all sorts of debris because there's such velocity inside them. We were flying along—around 1,000 feet in the air, and a log that looked about 15 feet long came out of that cloud over our heads, just past the rotor. By daylight the next morning, when we could first assess it, we had 7,000 acres burning. You could see the fire from over the mountains at Vernon. It was a monster. They tell me—I didn't see it—that the fire was so intense over so many miles of creek that the water was hot enough to kill the fish.

*Helicopter pilot Lloyd Siver photographed on Baffin Island in 1962.* Photograph courtesy Lloyd Siver

Morris: We went back later and took out the Cat that I'd been running. What was left of it. The fire melted the radiator right out of it. There was a power saw hanging on the back of it and you couldn't even tell what kind it was, just melted metal.

Hopkins: We did some pretty fancy firefighting. While the fellows all had to be pulled out that night, they were all back fighting fire the next day. Then we were blessed with air tankers and helicopters and we pulled in some excellent crews from elsewhere. The fire burned out the Cottonwood Creek basin, and I was afraid of it going over the ridge and raising hell in the Greenbush basin. We had a crew of smoke chasers at the top of the basin, which was mostly alpine. It had been such a dry summer the alpine was dried out, and it burned too. These fellows were up there with a little Homelite pump, running around all through this alpine, catching these smoking areas, little spot fires here and there. They had a helicopter for moving around, and they did an excellent job. They kept the fire from going over the ridge.

If one mystery remains, it involves the actions of Sig Ohashi during the near-lethal hours after the Hound jumped its guards. Two things are certain: Sig Ohashi's bulldozer burned and Sig Ohashi didn't. He's no longer alive to set the record straight, but the stories of his close call live on among those who were there. Fred Derry was the Crown Zellerbach logging superintendent

*Flames advance through slash on the periphery of the Hound fire.*
Lloyd Siver photograph

who worked as fire boss on the Hound. Derry recalled Ohashi being with a group of men trapped by the September 1 blow-up.

> Derry: When they got their wits together they went out the back way—through the top end of the fire. There were about 12 of those guys, and he was one of them. He thought he'd had the biscuit, so he buried his wallet and some other things in a wet spot. He went back later and got them.

Others involved with that fire have their own versions of the event.

Hopkins: One of the logging bosses, a Japanese fellow and a well-known logger, was running his Cat and he had to abandon it. There was a log landing there, so he cleared out as much as he could, dug a pit in the centre and put the Cat in it. Three loggers came by and said, "You'd better come with us. You're going to burn up here." They got out of there. The Cat was incinerated. So would he have been.

Hartley: One of the Ohashis, I think it was Sig, buried himself in a mudhole. There was a bit of a dip in the ground that contains water when the snow melts, and it never really dried out. He just dug a hole in a mud puddle and jumped right in 'er. We came along and found him there. So Sig's standing there within 50 feet of me, and that was his decision—to bury himself in the mud. It was kind of like everybody had their own thoughts, what they were going to do. We left him and I forgot about him even being there. The fire went right over top of him.

Morris: Kiyo [Kiyoshi Ohashi, Sig's brother] was working on the same road as us, but he was above us, and he managed to get up around the top and come back through where it had already burned. But I couldn't get up there because my swamper took off. He didn't stick around, he ran up to where Sig Ohashi was—that's one of the bosses. My swamper ran up to where Sig was because they had a pickup up there. They were on a bit of a landing. Siggy drove up there in his pickup, but he didn't drive it out. It burned right there. The fire went right over the top of them.

Kiyoshi Ohashi: There was a young guy there with my brother—a boy helping fight the fire. There was a kind of muddy place there, with water in it, so my brother said, "Let's dig a hole here—we can stay here." The boy said,

"No, we can't stay here. The fire will be too hot and too smoky. Let's go out where the fire has already burned. It's smoky there and pretty hot, but I think we can get out that way." And that's what they did. They walked out through the burnt area. It was still burning here and there, but they sneaked their way out. That boy probably saved my brother's life. The machines and stuff where they had been were all burned to a crisp.

Fire managers get ground down by the weight of making endless decisions that could influence their operation's success or failure. This was particularly true in the pre-union Forest Service, when staff could be called on to work unlimited overtime. Ranger Hopkins describes one result of this policy.

Hopkins: One of the hardest things I ever had to do with my staff was when I observed that they were suffering fatigue. When they did that, they made poor decisions and I had to get them off the fire. And every time I had to do it was a fight: "You don't think I can run this fire?!" It was *their* fire and *they* wanted to handle it. But when their eyes got that look and their voices started to slur, it was time they had a day off. I never found that anyone involved with a fire accepted that graciously.

In one of his reports from the 1938 Bloedel fire, journalist Torchy Anderson had described how forest ranger Charlie Langstroth was pulled from active duty after two weeks of frenetic activity. "Charlie could not remember when he'd went to sleep last," Anderson wrote, "but was loathe to quit the job when ordered." The Hound fire forced Vern Hopkins to make a similar and very personal fatigue-related decision.

*Logging contractor Sig Ohashi. How he survived remains one of the Hound fire's enduring mysteries.*
Photograph courtesy Kiyoshi Ohashi

Hopkins: I remember when I got back from my helicopter ride the evening that the fire took off, I landed and got in my car. The psychological effect of the fire blowing up like that absolutely surprised me. For weeks we'd been fighting fires in that drainage, first the initial four lightning fires, then this one that blew up. I remember driving away from the fire, then I had to pull over to the side of the road. And I bawled like a baby. That isn't my nature. But it was the buildup of tension, and the fact that I'd just gotten word that there were 12 men missing. And I really thought they were dead.

I called the office in Kamloops and said, "I need someone to take over here. I need a day off." It had been virtually a 24-hour-a-day thing, and three days of that is all a human being can stand. I was worried that I couldn't make decisions any more. So they brought down a guy from Williams Lake, Jim Ward, a very experienced and competent ranger, and he took over. Nobody was more relieved than me. I had a little boat down on Kalamalka Lake, so I took that boat out in the middle of the lake and went to sleep where no one could get at me.

Though he has fought other fires, it is the Hound that remains most clearly fixed in Garry Hartley's memory. "I never forget that fire. In a few seconds you could be dead if you didn't do things the right way. You do the wrong thing and you're not going to be around any more."

Former forest ranger Vern Hopkins had no difficulty explaining why the Hound, for all its explosiveness, did not become a fatal fire. "The important thing was that these were all industry men. Each person had their own foreman who knew the name of everybody there, so no one was abandoned. If we'd have been working with ordinary pick-up crews—people taken out of bars and off the street—we would have lost people for sure."

Hartley acknowledged that firefighting was about the last thing a logger wanted to do. His opinion was at least partially formed by the way the Forest Service treated him when the Hound fire was over. Fallers in those days provided their own chain saws, and Hartley had recently paid $350 for the saw that was destroyed when he and fellow firefighters ran for their lives during the September 1 blow-up.

> Hartley: In them days we had kids—we lived from pay-cheque to paycheque. So I had to go out and scratch for another chain saw, then wait at least five months to get paid for it. I'm dead against the Forestry for things like that. That wasn't the ranger sitting in his office, that was the government—the head boys doing that. Sig Ohashi had the same battle to get paid for his skidder and stuff.

There were other reasons that loggers dreaded fire season.

> Hartley: Basically a fire was a fire: a dirty job, dangerous. It was compulsory to go. In the sixties we were making 35, 40 bucks a day falling. Then we went and fought fire for a dollar and a quarter plus 50 cents for the saw—a dollar seventy-five an hour. So none of us appreciated doing it, but it was something we went and did because it was part of our livelihood. We had to protect it.
>
> I can remember one Friday night, we all come out of camp and stopped in Lumby for a beer. The next thing, in comes the Forestry and we're all in the trucks, going out fighting fire. That's the way it was in them days. Today they don't do things like that unless they're really desperate.
>
> Nowadays Forestry takes people and gives 'em a crash course and suddenly they're firefighters. I disagree with that. As far as I'm concerned they're not trained properly for it. You've got to be in the logging racket. When you're running from a fire, in your mind you've got to be able to tell what kind of ground you're headed for. There's places where you can hit windfalls that'd take you all day to get over. It's hard to describe, but I can glance through the bush and tell which way you should be going.
>
> All-round loggers are still the best ones to fight fire.

# — Fire and Ice —

After the Hound blaze was controlled, Lumby ranger Vern Hopkins had to deal with a series of lightning fires in the Greenbush Lake area. He had a camp established there for his fire crews but was frustrated by the toll the summer heat took on food supplies. Helicopter pilot Lloyd Siver, who was working for Hopkins that sweltering season, helped resolve the problem.

Siver: One day they brought in about 200 pounds of hamburger, and it all rotted before anyone could eat it. That was in the days before they had refrigeration in the camps. Hopkins was mad. He said, "I told those guys to dig a hole and put ice in it." This was what you did for refrigeration at the time. So Hopkins had me fly him up to Cranberry Glacier. He cut about 1,000 pounds of ice and I flew it back down to the camp. The firefighters dug a grave-sized hole, lined it with wool blankets and put the ice in there. That worked really well. They used to have nice little cans of apple juice in those days, and I put a bunch of cans in one of those holes for my crew. I came back about three weeks later and that juice was still cold.

In 1982, after Vern Hopkins had retired as fire prevention officer for the Smithers forest district, he learned that the Lumby ranger station, administrative headquarters for the forest district he had overseen for 12 years in midcareer, had been deemed surplus in a Forest Service restructuring program and was up for sale. An entrepreneurial spirit, Hopkins bought the place, converted the buildings to residential housing and resold it. He now lives in Osoyoos.

# CHAPTER 6

# The Ministry and the Industry

The relationship between the forest industry and the Ministry of Forests is one naturally fraught with the potential for conflict, an inevitable situation when people attempting to profit through resource extraction are overseen by bureaucratic authority. Until at least the 1980s, when the Forest Service began putting more emphasis on using seasonal suppression crews, BC's logging industry was an essential element of the province's firefighting force. Though loggers continue to be responsible for fighting fires that break out on their own operations, for decades they were the first place a ranger or his assistants turned when they needed good help fast.

Ernie Graffunder grew up in the North Thompson Valley, where he farmed, logged, and did contract packing with his own string of horses. During a stint of Forest Service employment in the late 1940s he once conscripted a sawmill crew who didn't get to go home for a month. Among them was his future father-in-law. "The best firefighters that there are is loggers," he said. "They know how to use their tools. They don't want the bush to burn down because it's their livelihood at stake. They know where all the roads are, the creeks and everything. If there's any water around, they know where to find it."

Though the ministry/industry relationship has always been in a sense symbiotic, in fire season the Forest Service has traditionally needed the loggers a lot more than the other way around. It was possible to push bush crews too hard, particularly in hot fire seasons. Diplomacy was therefore critical since, as former forest ranger John Weinard noted, "there was never a dollar available to buy them a drink some night in a bar—something to recognize their efforts. About all you could do was make sure their next timber sale got through as fast as you could humanly do it."

> Weinard: We did a lot of work with the forest industry, particularly during the time when they ran bush mills out in the woods. Out there you had—at the sawmill site—a cookhouse and a bunkhouse and 15 or 20 guys working. And these mills were scattered all over the place. And you had a bulldozer or two there. So these were ready-made heavy-duty suppression crews. *If* you could pre-organize it. That became the big thing to do; your effectiveness as a ranger was judged according to how well you could organize these entities before the fire season started. Some people did that very well, so they just sat and pushed the buttons when things got hot.
>
> If you didn't have good relations with everybody you worked with, you were dead in the water. We used to run awfully close to the line sometimes. You'd get a situation where you had a few fires and you needed some heavier duty people. The guy who was handiest to provide you with a crew was the guy who'd been out last time and the time before that. And this guy was basically interested in sawing trees into lumber.

Sometimes, in a bad year, tempers would wear pretty thin. You'd go to a mill owner and he'd say, "How about so and so across the valley? He hasn't been out for two years. Go and get *him*." Or everybody would say, "Look, it was a long breakup, we're behind here, we've got to get moving, we've got to make lumber. Why don't you pick up some of those guys around town?"

And you'd have to tell him, "You're right. We're tapping you guys too much." Then you'd go downtown on main street and into the beer parlours—anybody who's hanging around and not apparently doing anything. Maybe the odd rancher and his hands, as long as it wasn't haying season. That was always a big deal—"I'm not going anywhere. It's haying season."

In 1960, the season of the Dean, fire crews were drawn first from logging and sawmill operations, the people who knew the bush and were equipped for the work. As a second choice—when loggers were unavailable or already fighting fire—the Forest Service hired off the street. The problem, noted Don DeWitt, the man charged with subduing the Dean fire, was finding the right people to do the right jobs.

DeWitt: You had seven or eight good foremen lined up— fire bosses, camp foremen. But when you got a run of fires, you ran out of those men. So there were assistant rangers that you brought in from other districts. Some were good, some were learning—we were all learning. Then you'd pick up a crew. You had to look at a guy, size him up, ask him if he had any experience. Of course they all said they had experience—they got a few more dollars for running a power saw, running a Cat. So you picked. Sometimes you picked a good one, sometimes you picked a not-so-good. We accepted that. It was all we could do.

In an emergency the Forest Service resorted to the firefighting equivalent of the press gang—"blue-slipping," so called for the colour of the Orders to

Report forms. Bars were the most common source for these emergency recruits. A forest officer would walk in the front door—often with a police officer at his side—and ask who wanted to fight fire. If too few hands went up, anyone else was selected. (The distinction between volunteering and conscription was often a fine one, given that people who didn't offer to go when asked knew they would likely be forced into service anyway.) Sometimes bar patrons would clear out the back door as the local ranger came through the front; often another forest officer was waiting for them in the alley with a truck and a pad of Orders to Report. But "blue-slipping" wasn't resorted to as often as popular myth would have it, mostly because forest officers knew that beer drinkers yanked unceremoniously from their watering holes would at best provide minimal fire-line productivity.

> Weinard: It was used only in a dire emergency, where you had to have somebody and they weren't willing. You'd use it as a final straw. I've only used it a few times. One time was in Horsefly. There was a young fella in the area, in his middle 20s, an awful drinker. All the women around that country would die to get a second look from him. Hard-working guy, an Irishman. He worked in a sawmill.
>
> One time I needed men to send up to Mitchell River, a long ways away. This guy was stone drunk, and he said he wasn't going and his crew wasn't going. I said, "You've got to go, you're the only ones left here. Your crew is it. You're going to Mitchell River."
>
> "No, I'm not going. I've got a date tonight." This was pretty drunken talk. Then he said, "What are you doing to do, blue-slip me?" Everybody else was standing around to see what was going to happen.
>
> I said, "I hate to do that to a guy like you, Jack, but if it comes down to it I guess that's what we'll have to do."
>
> He said—and he was a big guy, too—"You and who else is going to blue-slip me?"
>
> I said, "Just me, Jack. I'm the one that's going to do it."
>
> So I took out the book, wrote it out. He's teetering away. I hand him the slip. He's studying the thing up close, trying

to focus on it. He said, "Well, if I've got to go, I guess I've got to go. But I won't work. You won't get a bit of work out of me."

So be it.

The crew rounded him up, poured him into the rig, and we take off down to the junction at Quesnel Lake. Get everybody into the boat, head up the lake, up the north arm and up Mitchell River. They got him unloaded and set up the camp. By the next day he was pretty well sober. I had to leave to organize for some other fires.

About four days later I went back to see what was going on up the Mitchell River. I got to the camp and my God, I couldn't believe it. It was heavy going—heavy timber on a steep slope. And there was a fireguard cut from the top of the mountain down to the river, for probably three-quarters of a mile. It was about six feet, eight feet wide. Cleared of everything. I was just amazed; I couldn't believe that much work had been done.

I got talking to the cook, and I found out that Jack, when he sobered up, was in such a savage mood that he just climbed on that crew, up one side and down the other. It was the best and the quickest fireguard through the toughest country that I've ever seen in my life. And he was quite proud of it. When he came down the mountain I got talking to him. He said he figured if he was going to be up there he might just as well make it worthwhile.

Fire-line rebellion isn't a phenomenon limited to days gone by. Art Hart ran heavy equipment for more than four decades, having begun earning pay on a D-4 Cat when he was 13. For most of the past 40 years he has lived and logged in Valemount.

Hart: The relationship with Forestry was good back in the old days. The Forestry guys or the fire boss would walk up to you on a fire and ask you what the hell to do. Everybody got along good. It's more snotty today. Some Forestry guys are good, some guys ain't. I chased a Forestry guy the hell out of here with a Pulaski [a combination axe and mattock] this past summer [1998]. He wanted me to cut a bunch of snags down. There was little fires burning under these bastards. You pile all these snags on top of these little spot fires, *now* you've got a fire going. And you ain't going to put it out. They're nuts. They had guys up there falling snags like a son of a bitch and got the fire really going.

This Forestry guy come down to where I was looking after my crew. He come down yapping at me about falling these snags. I told him, "You go up and look after your own son-of-a-bitchin' fire. I'm looking after mine." I said, "If you don't get out of here I'll split your goddamn head open."

As far as I'm concerned, the Forestry guys never worked. We had the same-sized fire as they had, and we had ours out three or four weeks before they did. I had all loggers down where I was working. As far as I'm concerned, them guys *worked*. They don't piss around.

One result of emergency recruitment was that forest officers at times became de facto social workers. When the Dean fire blew up and began rolling toward Merritt, busloads of recruits began arriving from the Okanagan. Once disembarked they became the responsibility of assistant ranger John Dearing, among others.

Dearing: Sorting out people was a big part of the job. You ended up running a social program. You had people who were a danger to themselves. They were intoxicated, frail, not properly dressed. And there they were—out on the fire line. Not only are you sympathetic and charitable towards them, but you have a responsibility for their well-being. Some people were truly unfortunate and weren't any good to you. Those ones you tried to get back to where they came from. Sometimes, when they fell off the bus, I'd just put them someplace where they could just sleep for a day or a day and a half. Then I'd feed them for a while, and they'd be okay. They were very nice people. Socially there's nothing wrong with them; they're just desperate people.

Recalling the treatment of the tin-canners during the Bloedel fire, Dearing noted that, "In those days, people of that sort—people without money—were treated like waste."

Whether volunteers or conscripts, green fire recruits received little in the way of preparation for the situation they were about to be thrust into. On the Dean, their initiation began with a pep talk. *Merritt Herald* reporter Pat Healy recorded the message ranger supervisor Don Fraser gave one group of new arrivals before sending them to the fire line.

> Fraser: Whatever you do while you're fighting fire, don't panic. Fire climbing green timber will go up with a roar. It will sound like hell breaking loose but it will stop just as fast as it went up. When you're working along the firebreak always keep the man on either side of you in sight. If you're trapped, go into the burnt-over area. That is the safest place. Now, we are going to hold this fire right here. If we don't stop it here, God knows where we will stop it. If we don't stop it here, Merritt is threatened. But I know that we can and will stop it. If a spark jumps over the firebreak, grab it with your hand if you have to. Sit on it, eat it, but don't let it get started on the other side of the firebreak. If it gets away on us here it is because you men are not doing your job.

Jim Bufano has fought fire as both a logger and a government employee. He served on an initial attack crew and later as an initial attack foreman during six years of Forest Service work in the Sayward area on Vancouver Island. A fire junkie—"It was an outstanding summer; we had fires all over the place," is his summary of the 1985 season—he began working for Weldwood in 1988. To his pleasure, firefighting was included in his job description.

> Bufano: In terms of responding to small fires, the ministry is outstanding. They teach you that the first three-quarters of an hour to an hour on a fire is critical. And that's true. As

an employee in that position I can say that we were very effective in limiting the size of fires.

From an industry standpoint, the ministry leaves a lot to be desired on larger fires. Lots of power struggles, including mini-power struggles between, say, mid-level industry engineering staff and mid-level ministry staff. They're fighting over things like radios. If you have a radio it means you have authority. Or fighting over who's in charge of which sector.

In a lot of instances you had guys standing around twiddling their thumbs because the ministry assigned too many people to the fire. It can become a morale problem. You get loggers, they want to do something. You get guys running around, bumping into each other—pretty soon they're sitting around, questioning what they're doing. Often using really colourful language.

Bufano's boss while he worked in private industry was Alex Fonagy, one of the group of forestry students who immigrated to Canada with their professors in 1957 when the Russian invasion of Hungary interrupted their studies at Sopron University. Fonagy and his fellow immigrants completed their training at the University of British Columbia. In 1962 Fonagy was hired as an axeman by Canadian Collieries, which later became part of Weldwood and then Interfor. He retired 33 years later as Interfor's manager of forest operations for Vancouver Island and part of BC's coastal mainland, primarily in Toba Inlet. He was ultimately responsible for fire suppression on the company's tree farms and, as he noted, "Sometimes I was responsible to fight the Forest Service on behalf of the company."

Fonagy: The company had to fight fires, because it held TFLs [tree farm licences]. You have the responsibility to maintain firefighting equipment. According to Forest Service regulations you are committed to fight fire with all your men. We didn't have to set up an organization when we had a fire—we had an organization right there. We had foremen, a superintendent, engineers, foresters. We could turn around and fight a fire on short notice.

Usually, if it was a big fire, the Forest Service took over. If they figured it was in their interest to take over the fire, then they took it over and they supervised it. They also took over certain costs. Until then, the company was responsible to fight the fire. We would supply most of the equipment as well as manpower, a cookhouse, that sort of thing. If they needed more people we supplied them, too.

Most of the time the Forest Service just sent in people to look over what we were doing. And we hated that. They didn't usually have much experience. You get a 20-year-old kid coming in and telling you what to do, you didn't like that. And then, of course, they reported everything back to the ranger. You always had this little fight going on between the industry and the ranger, arguments over various things.

Sometimes it was a case of safety. Some of the things they told you to do, if you did them you'd be dead. I remember one time in Toba Inlet I refused to send a crew into a certain place because I figured it was unsafe. Toba is steep. Forestry was controlling the fire at that time, and they wanted the crew to go in under the fire where it was *very* steep. They wanted to have a fire trail, a guard, at the bottom of the mountain. I refused to send the crew in there. Rocks were coming down, sometimes whole trees falling and sliding down, sliding 500 or 600 feet. Later on I sort of got into shit for that. I had to go to head office, and the Forest Service and I had a little discussion about why I refused to send the crew in.

Sometimes the government had the attitude to fight a fire no matter what. Industry didn't always agree. That was the basic argument between industry and the Forest Service: whether to sacrifice everything. Of course there was politics involved, too. It was a good point for the Forest Service if your district got a fire out fast.

And of course the Forest Service would send 100 guys for a little fire. You couldn't move sometimes, there was so many people. Money was no consideration whatsoever. The argument always seemed to come to this: they seemed to have an unlimited amount of money to throw at the fire. And of course, if they were coming to the end of their year and they hadn't spent all the money in their budget, then they'd *really* throw money around. You know the way it goes in government, if you don't spend it this year, you don't get as much in your budget next year. So then there were all kinds of helicopters, whether you needed them or not.

From the industry point of view, we wanted to get the fire out and go back to work as soon as possible. We also wanted to spend as little as possible—I'll admit that. But I'll say one thing, I always had a free hand in controlling fires. The company never told me that I can't use this or I can't do that because it cost too much money. Never.

# CHAPTER 7

# Paint it Red

Despite its vast forests, most of northern BC was, until the 1960s, so inaccessible that large-scale timber harvesting and processing were economically impractical. That situation only began to change when projects such as the creation of Williston Lake signalled the province's readiness to invest significantly in northern industrial infrastructure. Up to that time, if trees couldn't be profitably converted to currency, they were, in effect, largely valueless. So what was the point of spending good money to prevent them from burning?

Furthermore, obstacles to fighting northern conflagrations were overwhelming—as they can be even today. The region's remoteness was such that entire watersheds could burn unnoticed. Prior to the late 1950s, when helicopters began in a tentative way to provide easier back-country access, the only means of getting to and from most northern fires was by foot, horseback, riverboat and at times float-equipped aircraft; moving people and equipment in the quantities necessary to fight large fires amounted to a near impossibility. This wasn't necessarily a bad thing, since forest ecosystems evolve in the presence of fire and suffer from its oversuppression. A frequent challenge in the north was, and often still is, deciding which fires should be fought and how much money should be spent on them. Sending out a few guys on initial attack was one thing, but if that failed, the necessary commitment could be enormous.

# Paint it Red (The Tee Fire, 1971)

That challenge surfaced in a big way in July 1971 when lightning ignited a fire on Teeter Creek, 12 kilometres west of mile 496 on the Alaska Highway. The fire was thought to have started when an electrical storm swept through the area on July 19, though it was not discovered until the following day by a Forest Service patrol plane. The pilot reported that two lightning strikes were burning close together on a steep hillside, their total area approximately two hectares. By that evening, when assistant ranger Bob Rowsell did his initial helicopter reconnaissance, the fires had merged and grown to 80 hectares. That same evening a DC-3—an aircraft known in that country as "the Indian plane" for its frequent use in moving Native crews around the region—delivered 40 firefighters in two trips from Fort Nelson to the Liard Hotsprings airstrip.

There was a brief period when it looked as though Rowsell and his crew might extinguish the Tee. A very brief period. Firefighters got in nearly one whole day on the fire before things started going bad. On the afternoon of July 21 their sole helicopter broke down, stranding 17 men away from their camp overnight. The next morning an Okanagan Helicopters Sikorsky S-58T arrived, deposited another crew at the fire and recovered the stranded men. High winds then whipped up flames and prevented the Sikorsky from water-bombing with its monsoon bucket. A day later it began dousing hot spots but almost immediately developed mechanical problems.

The "T" in S-58T stood for turbo, and Okanagan had recently built—from a kit provided by the manufacturer—the world's first production model of that aircraft. Harvey Evans, the first person cleared to pilot the newly introduced machine, was two decades into a 42-year professional flying career and two years away from becoming the first person to fly a helicopter—the S-58T—from Canada to Greenland. The problem he was experiencing on the Tee was a repeated compressor stall which, considering the work he was engaged in, was a bit worrisome. "We were lowering the bucket right into the fire, then we'd dump our water, go to power up and bang, the compressor would stall. It scared the hell out of you because it was really loud and located right under your feet. We'd have enough power to get out of there, but it wasn't very good." Around the time that Evans stopped bucketing to have his engineering crew work on the compressor problem, a series of new fires sprang up around the one that Rowsell and his crew were fighting. The assistant ranger's notes for July 23 tell the tale.

0900—S-58 has not returned, must be having trouble.

1220—Another fire spotted across valley east of fire; approx. two acres in size and spreading rapidly up a hillside. Crowning.

1320—Fire across valley building up heat, spreading on all sides. Now approx. 200+ acres. S-58 has not returned yet. No contact with S-58.

1400—Fire across valley out of control and spreading rapidly. Smoke from another fire south of us and on bottom of ridge we are working. All crew called in to central helispot to enable fast evacuation if necessary.

1500—Fire east [of us] gaining ground rapidly—now about 500 acres. Fire to south billowing smoke; cannot see actual fire for an estimate of size.

1530—S-58 returned. Had mechanical difficulties with one engine which was stalling while water-bombing. Took crew to campsite to break camp. On way to camp spotted another fire in valley directly below ridge we were working on. This fire is about 250 acres and heading up valley towards camp.

1650—Camp broken and all men and equipment have been lifted out of valley to Mile 496. The fires are about to join one another and are really picking up steam. Phoned Fort Nelson to advise. Office staff to look into Cats, equipment and manpower.

1800—Ranger arrived at Mile 496. Fire now completely out of control and spreading rapidly.

Things deteriorated from there. The Sikorsky could work only sporadically as its compressor continued to falter. Four bulldozers were "walked" 19 kilometres cross-country to the mouth of the Deer River to begin cutting an east-flank fireguard. Catskinners were upriver the next afternoon, working about eight kilometres from the point where the Deer enters the Liard River, when

*Burning intensely downhill, the Tee takes a run at the Liard River.*
Bill Jackson photograph

the Tee made its first significant run—in their direction. Moving at ponderous top speed, the machines fled down the Deer and onto a gravel bar in the Liard, arriving there an hour and a half before the Tee roared up to the riverbank behind them. On its way the fire had burned through 9,000 hectares of "extremely fine merchantable timber" before moving on through the valley. Although a Forest Service report says only that "[The fire] travelled seven–eight miles in three hours," Ben Mitchell, the forester in charge of the Prince George region and a man who had seen a lot of big fires and was not prone to hyperbole, was clearly awed by what he witnessed from his helicopter. "This [timber]," he wrote, "was completely burned in three hours in the most spectacular fire I have ever seen ... It burned so rapidly in the Deer River basin that it consumed all of the available oxygen in gulps and was really a series of gigantic gas explosions."

This was the situation that Bill Jackson encountered when he arrived on July 26, the day following that big Deer River run. Jackson was a deputy

*A "pyronimbus" cloud formed by intense burning on the Tee fire begins to take on the typical anvil shape. Such clouds contain so many electrically charged smoke particles that they create their own weather. This convective column generated a lightning storm that touched off two new spot fires.* Bill Jackson photograph

ranger in Prince George at the time but was sent in to take charge of the Tee owing to his extensive fire experience in general and his knowledge of the province's extreme north in particular. Jackson had been bitten by the fire bug as a kid when his father, manager of the Penticton airport, torched the grass each year. In 14 years of Forest Service employment Jackson had worked around much of the province, including Fort Nelson. He had also learned some significant lessons about fire behaviour in 1966 when he helped clear the pondage area for the soon-to-appear Williston Lake. That wasn't fire-fighting, of course. It was fire setting—lighting up thick cured layers of forest that had been flattened by enormous crushing machines—but he credited that experience with greatly increasing his knowledge of and confidence around highly intense fires. When Jackson took over the Tee, he had at his disposal 11 Cats, four of them still stranded by heat and smoke on the gravel bar at the Deer–Liard confluence. The fire had jumped the Liard at that point, so he had a volatile spot fire demanding action on a difficult-to-access shore. He had about 100 firefighters—mostly Native people from Fort Nelson but also a few "hippie types" who showed up at the Liard Hotsprings camp looking for work—and two helicopters. And he had his orders.

Jackson: Henry Miskovich, the Fort Nelson ranger, decided that the only thing we could do was to get in as close as we could to the fire and keep it away from the highway and developments along the highway. The Liard River Hotsprings were right there on the highway, plus the Liard River Bridge was in danger as well as a few other bridges. The fire at this time was quite a few miles north of the highway, up Teeter Creek. We started gearing up, building Cat guards back into the mountains—we were building our own access, too. The fire kept making runs, adding maybe 1,000 acres a day.

*How do you cross a bridge with a D8 bulldozer? Very carefully. When the Tee jumped the Liard River near the Alaska Highway, fire crews shut the route down and "walked" this Cat across the bridge. The trip was agonizingly slow as a cushion of planks and tires had to be laid under the machine's tracks to prevent vibration from damaging the bridge. A truck caught in the roadblock was used to move three other D8s across. The Liard River Lodge at Liard Hotsprings is visible in the background.* Bill Jackson photograph

One of the "developments" the Forest Service wanted to keep the Tee away from was a major Canadian National Telecommunications microwave installation atop a hill at Mould Creek, near Liard River Hotsprings. The facility provided communication to the Yukon and much of Alaska and was a link in the continental defence system. An agreement with CN stipulated that a penalty of $3,000 per minute—$4,320,000 per day—would be levied against the company if transmissions were interrupted. If the installation were roasted it would clearly be down for quite some time, and no one wanted to do the math on that prospect.

On July 27 Jim Phelan, the head of CN Telecommunications in the Yukon, contacted Yukon forestry superintendent Edo Nyland to ask for help in saving the tower. Nyland and his staff had been tested by some difficult fires during 1970 but had been pleased with the results obtained from the use of an innovative portable mixing system. It made possible the on-scene preparation of a long-term retardant called PhosChek—a slurry of water,

*Flames are visible through smoke as the Tee rampages toward a critical CN Telecommunications microwave tower at Mould Creek on July 29, 1971.*
Bill Jackson photograph

phosphate-based fertilizer and dye, which allowed pilots to see where their drops had landed—and its subsequent application by bucket-equipped helicopters. After learning of Phelan's plight, Nyland contacted Harvey Evans, who had worked with the mixing system on the Yukon's 1970 fire bust. When Evans heard what Nyland and Phelan were trying to accomplish, he agreed that "water ain't going to do it."

Bill Jackson was also well aware of the microwave facility and, suspecting that the Tee was building up for another big run, had sent a pair of D8 bulldozers up the tower hill to scrape off everything burnable over as wide an area as they could reach. He had them windrow debris at the base of the hill and cover the debris with dirt, hoping that the resulting wall would deflect heat straight up, instead of toward the tower. The Cats were continuing with this project late in the evening of July 28 when the mixing station—which had been flown from the Yukon to Liard Hotsprings and set up at the mouth of Teeter Creek—became operational. One supply of PhosChek had arrived with the mixing unit, but Nyland had a second load flown in from Vancouver aboard a chartered Hercules transport plane. All this action impressed Harvey Evans.

> Evans: I thought, Geez, this is getting wild around here. I said, "How much retardant do you want us to put up?" They said, "All you can put up there. Paint everything red." We flew at it from dawn to dark. The money was mounting up, but expense didn't seem to mean anything. Just go. We literally painted the top of that mountain red. It wasn't until after the fire was over that we found out why. It was the last link between the United States and the DEW Line. The eastern DEW Line link was down for repairs and the coastal link was down for repairs. This was the only link left for the defence of the whole North American continent. And it was the height of the Cold War. So they did *not* want that tower to burn down.

Evans flew until nearly midnight, laying a series of retardant strips below the tower across the fire's anticipated path. It was at about this time that Jackson received a weather forecast for the following day—29.5 degrees Celsius and

*An innovative station for mixing long-term fire retardant was shipped in from Whitehorse on July 28 to help save the threatened microwave tower at Mould Creek. Here a ground crew pumps retardant into the bucket of the Sikorsky S-58T being flown by Okanagan Helicopters pilot Harvey Evans.* Bill Jackson photograph

72-kilometre-per-hour winds. Some serious fire activity was to be expected. On the morning of July 29 catskinners and crews along the Deer River worked for only an hour before being evacuated to the same Liard River gravel bar to which they had fled on the day of the Tee's first big run. Evans applied yet more retardant around the microwave tower and in the path of a new spot fire threatening to outflank defences set up at the base of the tower hill. Then the PhosChek ran out. The temperature was well on its way to a daily high of 33.5 degrees Celsius, winds were blowing at close to their forecast strength and the Tee was beginning to really cut loose. When it did, Bill Jackson said, "That fire blew like you've never seen a fire blow. At its peak there was a 36-mile flame front rolling up hills and down hills like a giant steam roller. I'd never seen anything like it in my life."

Jackson had a good view of the Tee letting go because he was on the hill at Mould Creek when it happened, determined to do his best to save the

*On its most volatile days the fire burned roughly 32,500 hectares of forest.*
Bill Jackson photograph

tower. However, in attempting to do so he found that he first had to give CN Telecommunications employees an introductory course on some elementary but very important firefighting techniques.

> Jackson: We had a CN Tel technician with us on the hill, and I was on the phone talking to the CN guys in Whitehorse. I explained to them that we wanted to burn off from up there. As soon as the winds started drawing properly—when the main fire began drawing air—we wanted to torch off that perimeter. The two fires would draw together and keep the fire and heat away from the tower. This is standard firefighting practice.

It may have been standard for Jackson, but it wasn't for Jim Phelan at CN Telecommunications, who was not about to approve someone setting a fire at the base of the hill atop which stood his $3 million tower. Furthermore, Edo Nyland was sitting in that office with Phelan and it was Nyland's contention, based on the opinion of a Yukon forest ranger he'd sent to the Tee to oversee the mixing station, that a burn-off would not be effective in the prevailing conditions. And whether it was standard firefighting practice or not, Bill Jackson had no intention of lighting that match himself. So what ensued was a sort of long-distance staring contest. Things were starting to get pretty hot as this was going on.

> Jackson: I spent an hour and a half, two hours, talking on the phone with the guys in Whitehorse. The problem was that we had to let the CN technician light the first match— so it would be the company's fire and their responsibility. I gave them my best talk about firefighting strategy, trying to convince them that this is what has got to be done. Otherwise, I said, we can bear no responsibility for what is going to happen. I said, "You've got a wall of flame coming that's 300 feet high."
>
> And they wouldn't go for it. The technician on the ground could see what had to be done, but we couldn't convince those guys in Whitehorse. An act of God—they could take that. But they didn't want to take any responsibility for lighting a match.

*Bill Jackson: "That fire blew like you've never seen a fire blow."*
Photograph courtesy Bill Jackson

The fire front was approximately 1.6 kilometres from the tower when Jackson's helicopter lifted him and the CN technician to safety. Fifteen minutes later the hill disappeared in a gigantic pall of smoke and flame. "Oh God," Jackson groaned, "there it goes."

As the Tee rolled up to and around the hill at Mould Creek, a nervous Jim Phelan was listening on a dedicated phone line connecting him directly to the microwave tower. On another phone he had an American general concerned about the facility's capacity to continue its uninterrupted role in defending North America. The funny thing, Edo Nyland

# — *The Victoria Outlook* —

In 1955 the territory that forest ranger Sterling Cosens was responsible for consisted of pretty much the entire northern third of British Columbia, an area that could swallow sizable European nations. His staff consisted of four far-flung assistant rangers, a dispatcher, a seasonal fire patrolman and two lookouts. Having grown up on his father's homestead near Pouce Coupe, Cosens remembered the sky being filled with smoke for weeks on end.

> Cosens: We had some gigantic fires in the Peace. The attitude towards the Peace River country was that there was nothing there—log and get it out, forget about what burned. That was the Victoria outlook on things. They had a policy, when I became ranger there in 1955, of fighting a fire if it was within one mile of the Alaska Highway and if it could be controlled with a crew of 10 men or less. It was a convenient policy that allowed them to do nothing outside of that strip. It was not very good practice. For one thing, when you got large fires there was so much smoke you couldn't find the small ones. So the small fires became big fires.
>
> That type of firefighting philosophy caused probably the greatest disaster in the North Peace, a fire in 1950 that took off and burned right into Alberta. People died in that fire—some Indians burned right where they lived. Once that fire got away it was long gone. It was a huge fire. BC's chief forester was in Finland at the time. He woke up in the morning and said, "Where is all the smoke coming from?" They told him it was from a fire in the Peace River country of British Columbia.

The fire Cosens referred to was the Whisp, which began in late June in the "free-burn" area north of Fort St. John and over several weeks spread into Alberta. In late September gale-force winds struck, blowing the blaze toward its final size of 1.4 million hectares—North America's largest recorded fire.

*A massive smoke cloud envelops the microwave facility. "Oh God, there it goes," were Bill Jackson's words at this moment. Later that day it was revealed that defensive work done at the site allowed the tower to survive with minimal damage.* Bill Jackson photograph

later commented, was that for all the tower's importance CN had done nothing to protect the cable that ran along the ground from the main power plant on the Alaska Highway up onto the hill. Burying it under the least bit of dirt would have done the job. As it turned out, between the Cat work that Jackson had ordered done on the hill and the retardant that Harvey Evans had plastered around it, the Tee's radiant heat warped one big satellite dish somewhat but didn't do significant damage. However, flames did reach the unprotected power cable. There was a pause about seven seconds long at that point, a pretty quiet seven seconds back in Whitehorse. Then a back-up power unit on the hill kicked in; Jim Phelan began breathing again and Edo Nyland smiled and the American general was able to rest easy believing that North America was still secure. And Bill Jackson could say, as he did many years later, "That was very gratifying. What we did worked."

As near as Jackson could figure it, the Tee had covered roughly 53,000 hectares by the evening of July 28. Twenty-four hours later—after the fire had taken its run at the Mould Creek hill, carried on toward the Alaska Highway and blown out toward every other point on the compass—it was about 32,000 hectares larger. There was a good news/bad news aspect to all of this. The bad news was that the fire now threatened the highway and necessitated the evacuation of Liard Hotsprings. The good news was the Tee had burned so close to the highway it was within easy reach of bulldozers; it had come to the firefighters. Furthermore, as crowning flames approached the highway they entered a band of deciduous trees—a less volatile fuel source—so they slowed and dropped to the ground. Cooler temperatures the next morning allowed Cats to begin cutting guards across hillsides above the highway and crews to burn off between those guards and the fire. In fits and starts the Tee continued to increase toward a final burned area totalling just under 110,500 hectares, but the bulk of that increase took place on the fire's hinterland perimeter, and it would never regain the volatility seen during the last week of July.

More than $350,000 had been spent fighting the Tee by the time the last smoke was seen in mid-October. Victoria was not amused; an explanation was demanded. Protection branch director Cy Phillips explained to superiors that it was not feasible to abandon the fire owing to the presence of the CN tower and the tourist destination of Liard Hotsprings. Considering the facilities at risk, "a slightly stepped-up token action was laid on," he wrote. In a memorandum to the province's assistant chief forester, Phillips implied that the Tee had not been as well managed as it could have been. Such remote suppression action, he said, "is difficult to supervise, particularly when it is over 200 miles from the ranger station at Fort St. John." In a terse memo to Prince George regional forester Ben Mitchell, Phillips noted that among other shortcomings on the Tee, "It is also evident that there was a labour problem including liquor difficulties."

Bill Jackson had noted as early as August 3 that his crews were beginning to show "signs of battle fatigue" and he suspected that they were ready for some kind of blow-out. By mid-August, with the Tee well into the drudgery of mop-up, most firefighters had put in 14-hour days for three weeks without a break. Deciding that it was time for a party, they had someone deliver them a shipment of booze from Fort Nelson. Jackson didn't raise any hell about that. Instead he poured the men onto a bus the next morning and sent them home. He was frustrated, though, when he found that the replacement crew arriving on that bus had smuggled a few bottles on board to shorten the long trip north. Some were "pretty gunned up" and a few had already passed out by the time the bus doors swung open. "The [Forest Service] boys would go in and drag them out by their heels," Jackson later commented, "their heads bouncing down the steps and onto the ground. Then they just laid them out face-up in the sun."

Mitchell, in his response to Phillips' criticisms, noted that the liquor problem was not unusual. "It is impossible to keep all liquor out of a large, accessible fire camp for three weeks, and the fire boss has to occasionally tolerate a group of Natives obtaining a supply and requiring a 'drying out' day. The only alternative is to let them go to town every week or so and this takes longer and is more expensive."

It is worth noting that Mitchell was revered by many ground-level Forest Service personnel for his intelligence and his fearless defence of the people in his command—"a prince of a man" is how one former colleague remembered him.

*Visible behind a curtain of smoke, a massive flame front approaches the Alaska Highway.* Bill Jackson photograph

# — *Riding the Indian Plane* —

Jim Supernault began fighting fire as a teenager in Alberta in the 1930s when the pay was 15 cents an hour and you supplied your own bedding. In 1964 he moved to BC's Peace River country where, like many Métis people in that area, he developed a reputation as a first-rate firefighter and worked his way up to crew boss—as high as you could get, he noted, without the benefit of formal education. In the mid-1980s his crew spent a month fighting a large, remote fire outside Atlin. As usual, their mode of transport was the venerable DC-3, "the Indian plane."

Supernault: When we were going in there in the DC-3 one of the motors quit. We keep flying, then finally the pilot said that we might have to throw some stuff out. The motor was pulling too much of a load. He kept radioing to a little settlement there, Ingenika Point [on Williston Lake]. There's only a little airstrip there, but it's the only chance you've got. You've got to take that chance. So we did turn around and come back to that one. When we were going to land there they told us what to do. We had to bend over and hold tight. Belts and everything. We had one little crazy guy there. He don't stop for nothing. He don't care. He said, "Spread your legs and kiss your ass goodbye."

When we hit land we kind of bounced and kind of tipped a little bit. The ends of the wings were pushing down the little trees on the sides. But we made it. They got that plane out and came with another DC-3. When we fly out we had to fly out over the lake. The airstrip came right up to the lake. It seemed like that son-of-a-gun wasn't going to leave the ground. Heading toward the water, it kind of scared me. But we left it.

Jim Supernault retired from firefighting at age 68. He died in 1999.

In plain, uncompromising language, Mitchell took full responsibility for the actions of the people in charge of the Tee operation. He described the tremendous logistical complications involved with the fire, as well as the extreme burning conditions. Explaining why he elected to continue the Tee fight even after the area's most valuable timber had burned—the 9,000-hectare stand he had seen incinerated along the Deer River—Mitchell turned the tables on his Victoria masters. The fire, he said, was by that point being recorded by a CBC television crew as well as by a BBC team that had diverted from making a documentary film nearby. "In addition to this," he wrote, "it was now only six weeks away from the day the premier, several ministers and the attendant press were to be in Fort Nelson to open the PGE [Pacific Great Eastern railway, now BC Rail] extension. I was faced with a socio-political decision regarding the wisdom of abandoning the effort and risking incurring public comment against the government for spending millions of dollars opening up northern resources with a railroad on one hand and saving some thousands of dollars by letting it burn up on the other hand." Responding to Phillips' recommendation that a post-mortem should be conducted on the Tee, Mitchell noted that since he was the person who approved all decisions taken on the fire, a post-mortem would ultimately result in him passing judgement on himself. "As you suggest in your memo," Mitchell concluded in his message to Phillips, "there is a need to update our thinking on firefighting in the north. We like to think that we are doing this on a continuous basis...It would take several pages of discussion to completely outline the variety of problems and values in the north, but we would be more than pleased to sit down and discuss some of our unique problems at any time."

# CHAPTER 8

# Buzz: The View From the Cat

Some know them as bulldozers, others as crawler tractors. People who work with them have honoured the company that made them famous—the Caterpillar Tractor Company of Peoria, Illinois—by referring to them universally as Cats. Those who run them are known as catskinners, and when it comes to firefighting there is indeed more than one way to skin a Cat.

For all their lumbering power and size—some are big enough that helicopter pilots use their canopies as helipads when surrounding ground offers a poor alternative—applying them to firefighting is a subtle art and a skill much esteemed by the people who depend on them. Some old-time firefighters consider them the most significant weapon in their arsenal, even more important than helicopters and water-bombing aircraft.

"There are catskinners and there are people who drive Cats." This sentiment in various forms is often expressed by fire bosses and their lieutenants. A Cat handler may be unequalled at flatland road building, for example, but next to useless on the extreme grades and hostile terrain that is common fare on fires. "Good catskinners on fires are hard to come by," acknowledged Tom Blom, a former fire boss who spent five years as BC's superintendent of fire control. He provided a job description for firefighting Cat operators. "You've got to have a catskinner that's a little bit crazy. A little crazy, a little cautious. Someone who'll do the job the way you want it done. There's not too many of 'em."

By the time he was six, Buzz Kurjata was using a little two-cylinder bulldozer to clear snow from rural driveways in his hometown of Chetwynd. As a teenager he cleared land after school and on weekends; for the past 28 years he's been running Cats full time. In 1985, when many operators were losing their heavy machinery to 24-percent interest rates, Kurjata stayed afloat largely by being the first catskinner to arrive at and the last one to leave the long-running Ran fire. Started by a lightning strike in mountainous terrain between Moberly Lake and Hudson's Hope, the 8,000-hectare Ran demonstrated the dangers inherent in that work when it claimed the life of a fellow bulldozer driver.

Although Kurjata has built roads, logged and cleared agricultural land, all from the seat of his Cat D7-G, the oil patch is now his bread and butter.

*Buzz Kurjata with his D7-G Cat. He ran his first bulldozer at age six and hasn't looked back.* Photograph courtesy Buzz Kurjata

When not occupied there, he puts the machine to use on the 850-hectare ranch he is developing northeast of Chetwynd. Any top-notch catskinner eventually attracts the attention of people responsible for fighting fires. Kurjata has been drawing that kind of notice for much of his working career.

Kurjata: I think the first fire I was on was out behind my father-in-law's place on the Jackfish Lake road, outside of Chetwynd. It was just a small bushfire, but it was in the spring of the year when the frost was coming out of the ground. Not being real experienced, I went down over a hill and got stuck. Needed a winch to get out of there. Of course, the fire was boilin' up towards where the Cat and I were sitting. I got a bit excited, so I flogged my T-shirt in the wind when Harvey [Evans] come along with the chopper. He phoned and got another Cat to come and winch me out; he sent the cavalry to the rescue.

Most of the time you're closer to the fire than you want to be. You've got all that smoke and you can't see 'cause your eyes are watering. It's hot and you can hardly breathe. You're wondering whether the fire's going to get ahead of you or cross the guard and start burning up the sides of you. Where do you turn to? Where do you run?

Most of the time the actual width of the fire—the face that's burning—is not that wide. With the Cat, if the fire's not too hot you can always turn into it and get in behind, where it's already burnt. But you're always looking for a place to run to, or else you're looking for that helicopter in the air.

When you're on the Cat, your world, what you can see, is about 20 feet around you, so you have to have the helicopter in the air, preferably every half-hour, so he can talk to the Cat foreman. They have to watch. If the fire jumps the guard down below us and we're halfway up a hill, the fire can come up that hill in a matter of seconds, and you're toast if it does. A fire can run up a mountainside in a couple of minutes. You can't go very far on a Cat in two minutes.

*The view from the driver's seat: "Most of the time you're closer to the fire than you want to be," says Kurjata. "Where do you turn? Where do you run?"*
Buzz Kurjata photograph

So it's really important that the helicopter's up there bird-dogging, watching things. Sometimes you'll get an inexperienced fire boss and he'll get worried about the cost. Helicopters cost lots of money to be flying, so they'll encourage the helicopter guys to be sitting on the ground. When they do that, you can get in a real bind. As long as you've got an experienced fire boss and he's not afraid to tell Forestry, "To hell with the cost," you can be pretty safe on the fire. But there's places that's just too dangerous to go without that eye in the sky.

On a Cat you can go just about straight down or straight up. On the Dunleavy fire—this was outside of Hudson's Hope in about 1986—it was the only time that I ever had to have a power saw man going ahead of me. It was so steep going up I could walk up with the Cat but I couldn't push any trees down. They were just little trees, but if I come up against them I'd spin out. A power-saw man cut the trees and felled them crossways and I'd crawl over top. I got up to the top that way, then I winched up the bigger Cat. Then we turned around and made one pass down to build the fireguard. I think it took us five hours to get up on top and about five minutes to go back down.

We cut guard down a real steep sidehill, one Cat behind the other. It was so steep we couldn't back up, couldn't turn around. We come to a creek at the bottom, the problem being that the bank was sloughed off just about straight for about 50 feet down to the creek bottom. We just took a big bladeful of dirt and jumped over the bank with it, down into the creek. We couldn't go back up the slope, so we did a real environmentally friendly thing: we went down the creek for about a mile and a half till we come to the bridge where the fire camp was. That's all we could do. It was straight up on both sides.

Usually the food in a fire camp is tremendous. In this area they hire a lot of Native cooks. The food is exceptional and there's lots of it. On some of the little fires they fly over with the chopper and kick out a box of Dixie Lee chicken and a can of pop. The pop hits the ground and splat! you've got nothing to drink. The Dixie Lee's got so much spice in it that it gives you heartburn. If you're experienced like I am, you've got a tin of Rolaids in your toolbox on the Cat. If you're not experienced, you eat as little as you can and have heartburn.

I was a Cat boss on a fire once for a day and there wasn't a water hole within a half a mile of us. We'd been cutting guard up this hill all day long and were pretty dried out, so I got ahold of the helicopter and asked if he could drop down half a dozen pop for me and the other catskinners. The first half a dozen he dropped, we got one tin out of that. The other ones all splattered. They tried again, and the second time we got two more tins out of six. The third drop they kicked out we got two more that didn't splatter, so the five of us each had a pop.

Some of us spent the night out on a fire at Lone Prairie. They'd sent in what they call an O-team [a group of specialists known as an overhead team]. I call it an A-team because it's a bunch of assholes, a bunch of egomaniacs. They're hepped up and they think it's a good career move with the Forestry to go out and be a fire boss. It got to the point where I told Forestry that I'd go out on a fire if [local Dawson Creek Forest Service personnel] were handling the fire, but that if that A-team showed up, I'd be on the lowbed going home. They're a horrible bunch to work for. They're not safe. They're not experienced. They left us out overnight on one fire and then they didn't want to pay the Cats 'cause we only worked till dark—which is about midnight. Then we started again when it was daylight—which is about four in the morning.

There was four of us that got left out. We spent the night cold, sleeping on the hood of the Cat. There's a little bit of warmth when you shut the motor off. Not enough. You don't really sleep, you just lay there and shiver.

When you're left out at night, you've usually worked all day. Even if you take a big lunch it's gone by six o'clock or so. You expect that by nine-thirty, ten o'clock you'll be into camp. So everybody was out of food, out of coffee. Of course, you don't really want to light a campfire to keep warm—fire hazard's a little high. But it makes you get up and go to work in the morning.

Before [the mid-1990s, when firefighting was made a specialty operation within the Forest Service], fighting fire was sort of everybody's job, and everybody had experience. Now very little experience shows up on the management end at fires. For example, they light their backfires on the wrong side of the fireguard. That's happened.

With the new Forestry regulations they're only allowed to work certain hours, and they don't like to pay them overtime. So you'll have a pretty good fire boss, someone who knows what's going on. And then all of a sudden bang! he's gone 'cause he's houred out. And some younger person'll come in that's hardly even seen a fire before. And now they're the fire boss. *You're* trying to do things safely, and *they're* more worried about the economics, 'cause they're afraid that if they go over budget they're going to get a bad name with the Forestry.

You try to tell those people that they're doing something wrong, but it doesn't generally sink in. They're the boss and they know what they're doing and you're just a dumb catskinner, so get the hell back on your machine and do what you're told. It's like they say, a good catskinner gets paid from the neck down. But it's made for some hard feelings. After one run-in with an O-team I said I wasn't going out on any more fires. But you can't let the country burn up, so I guess I probably will go. Hopefully it will be local people that are running it. Maybe even somebody with some experience.

I was the first Cat at the Ran fire and the last one to leave. I was there for 42 days straight. Lots of 20-, 22-hour days. I think I was on 24-hour days the first three days I was there.

That's the fire where Willie Hauber got killed. I was working with him that day. It was up in the mountains— steep country. Every day where you go to work in the mountains you're risking your life. I still say that you shouldn't use a D8 in the mountains. They're too heavy.

Willie was driving a D8, and he had just sidecut up over a rock ledge that was buried about two feet in the ground. When the inside track overbalanced on this rock ledge it threw all the weight of the Cat onto the sprocket on the outside. Being that we were sidecutting uphill and it was loose dirt, the bank just gave way underneath him and the Cat rolled. I think he was thrown off; I don't think he had his seatbelt on. The canopy flattened out on his Cat. His neck was broken.

It wasn't really anybody's fault, it's just one of them things that happens. It gives us all a wake-up call and makes us all be a little safer, snug our seatbelts up a little tighter.

Everybody took the next day off. I don't think anybody worked. And then we shut down again for his funeral. The fire could do whatever it wanted. Willie was very well liked and very experienced. It was the least we could do for him.

*Willie Hauber was killed while fighting the 1985 Ran fire. Fellow catskinners shut down their machines the day he died and on the day of his funeral. "It was the least we could do for him," commented Buzz Kurjata, who was working with him at the time of the accident.* Photograph courtesy the Hauber family

One thing that happened on the Ran fire seems a little humorous now, though it wasn't too funny at the time. The fire took a run up a sidehill that had been logged off. The Cat foreman told me to high-blade up to the top of the hill and get a guard around the front of the fire and then cut guard back down. I took off through all these big high willows and stuff. With the leaves on the trees I couldn't really see anything. I crossed a skid trail and came out right beside three prime firefighters hired straight out of the bar. They'd snuck away from the fireguard and were hiding on this trail, smoking a joint. If I had come out on that skid trail a Cat-width farther over I would have run over all three of them and never even knew it. They had no idea that they'd just come a few feet from being dead. They just gave me a big grin and a wave. That must be wonderful stuff.

# CHAPTER 9

# Eg-splosion

To appreciate what Ken Brahniuk was looking at on the afternoon of July 29, 1982, you've got to have a sense of proportion, a sense of the scale involved. David versus Goliath doesn't begin to touch it. A boy with a sling and a pebble would have felt positively well-armed compared to Brahniuk, equipped as he was with nothing but an ailing helicopter. Doing battle wasn't a consideration for the time being. Flight was what he had in mind, that and an accompanying sense of mortality as a wall of flame rolled toward him across the vast muskeg of the Liard Plain.

The day hadn't started so dramatically. Brahniuk and his crew of four dozen firefighters had finally been getting a handle on their assignment, the Cran fire, along the Liard River. Getting a handle on a wildfire was an accomplishment that summer, when northern BC and the adjacent Yukon were struggling with an outburst of major blazes, some of them so large, so remote, or so smoked-in that their perimeters remained uncharted. Just across the 60th parallel, the Yukon Forest Service struggled with Watson 12, which was on its way to a final area of nearly 120,000 hectares, and Watson 13. To deal with its share of the bust, the BC Forest Service had established a command centre at the border community of Lower Post, only to have that town's dump fire rage out of control to the tune of 3,150 hectares. Here and there around the province's north, big, troublesome blazes—among them the East, the Pine, the Hall, the Frog—ensured that no one in the fire business was idle in the summer of '82.

## Eg-splosion (The Eg Fire, 1982)

At the time of the outbreak, Brahniuk was responsible for forest-fire protection and recreation programs in the Queen Charlotte Islands. On July 24 he got a typical Forest Service drop-whatever-you're-doing emergency summons. "Someone phoned me up and said, 'I don't care what it takes to get you here, we want you tomorrow. If you have to charter a Lear jet, do it.' I chartered a Beaver. That's all I could get. I was almost deaf from the roar of that thing by the time I got to Dease Lake. Then I caught a helicopter to the Cran fire."

Though Cranberry Rapids is actually on the Liard, roughly 30 kilometres southeast of the fire, someone had chosen it as the closest significant geographical feature to use in naming the blaze. The tiny settlement of Fireside was located adjacent to Cranberry Rapids, but the Forest Service did not use community names when identifying fires. That policy aside, Fireside would soon enough be closely associated with wildfire.

When Brahniuk arrived on the Cran, the fire boss he was replacing gave him a cursory introduction to the situation and, before departing in his heli-copter, handed over a Players Light package on which he'd listed the human and mechanical resources at Brahniuk's disposal. (Brahniuk, noting how exhausted the outgoing fire boss was, assumed he was being sent home for a rest. In fact, Tim Van Horlick was only being reassigned and was well on his way to racking up 404 hours of on-duty time for the month of July.)

Compared to other conflagrations in the extended vicinity, the 250-hectare Cran was a campfire, but its size belied its challenges. Much of it was burning in steeply gullied terrain running down to the Liard. The area's fine-textured soils plagued Cat operators on steeper slopes where their tracks

*Forest Officer Ken Brahniuk.*
Photograph courtesy Ken Brahniuk

*A crew of firefighters await deployment to Eg trouble spots.*
Douglas Cowell photograph, BC Forest Service collection

churned into the stuff as though it was so much sugar. Though the Cran was being fought by 30 skilled Native firefighters from Lower Post, when Brahniuk dropped out of the sky and became responsible for putting it out, it was uncontained by fireguards.

Brahniuk and his crew were backed up by Cats, water trucks, tank-equipped skidders and a big Bell 214 helicopter rigged with a 2,700-litre water bucket. With the Liard serving as one of the Cran's boundaries, the 214 had a rapid turn-around time. Repeatedly dipping loads of river water and dumping them on the hottest spots, it played a crucial role in squelching the volatile flare-ups that often threatened to wrest the blaze further out of the crew's control. More assistance arrived when Smithers-based forest officer Harry Hutchinson joined Brahniuk as his second-in-command.

On the afternoon of July 27 Brahniuk was doing an aerial reconnaissance in his little Hughes 500-D when he saw that flames that had burst across a fireguard had surrounded a Cat and its operator. As the 214 swooped in to drown the flare-up with bucketfuls of the Liard, Brahniuk's pilot touched down on the Cat's canopy, allowing the driver to scramble aboard and be lifted to safety. With such graphic evidence of the Cran's volatility, Brahniuk

*Cutting fireguard with Cats.* Douglas Cowell photograph, BC Forest Service collection

*A bulldozer operator works in intense heat as he widens an Eg fireguard.*
Douglas Cowell photograph, BC Forest Service collection

increased his crew size to 48 and maintained a steady night patrol, and by the evening of July 28 he considered the fire under control.

The next morning dawned remarkably calm and stiflingly hot. Windless mornings on fires are generally a cause for celebration, but somehow optimism was not Brahniuk's first impulse when he climbed out of his tent. What struck him instead was "a kind of eerie calm; I had the feeling that something wasn't quite right." The thing that wasn't quite right was something over which Brahniuk had even less control than he had over the Cran fire. What wasn't right was the Eg.

More than a month before a lightning strike touched off the Cran, a BC Forest Service air attack officer—one of the "bird dog" officers who lead air tankers to their targets and dictate their drop strategy—had witnessed two lightning strikes as he and his bomber group lifted off the runway at Watson Lake. Diverted from their intended target, the group flew to the remote Egnell Lakes but discovered that the fire had already grown so large and was burning so aggressively that it was "beyond resources"—that is, there was nothing anyone could immediately do to fight it.

Located near the heart of the sprawling Liard Plain, the Eg fire epitomized the challenges firefighters face in BC's north. In that area no roads existed and the Liard River was unbridged. The Forest Service considered barging in heavy equipment but discarded the idea as impractical. Then, having learned from Cassiar Forest District managers that the fire was burning in an area of low timber value, and having other blazes demanding attention, Paul Pashnik, the Forest Service's man responsible for fire suppression in the region, elected to let it burn largely unfought. After all, the Eg was surrounded to the west, north and east by the Liard River, a natural control line broader than any Cat could cut. If at some point the blaze threatened specific values worth fighting for or if natural features allowed for strategic backfiring, a localized suppression effort would be considered. In the meantime, it would do little more than create another expanse of scorched muskeg and, in doing so, would contribute to the maintenance of the region's fire-dependent spruce-pine ecosystem.

When he took over the Cran, Brahniuk had been given the additional responsibility of keeping an eye on the Eg's eastern flank, which he did daily, flying close enough to verify its progress. On July 25 the Eg and Cran were separated by about 25 to 35 kilometres, by Brahniuk's estimate. Though this margin of safety appeared substantial, he recommended that the Dease Lake Forest Service office advise residents of Fireside and nearby Coal River that a potentially dangerous giant was gradually—though not yet inexorably—moving their way.

At about this time a thermal inversion clamped down over the Liard Plain, a vast atmospheric lid that produced stifling, almost windless conditions. Effective helicopter surveillance was at times impossible owing to the layer of trapped low-level smoke. Tim Van Horlick noted that the claustrophobic conditions were making people irritable. Driving from his base at Lower Post to the border community of Contact Creek, he found that wildlife in the area also seemed to be reacting to some invisible force.

> Van Horlick: As I drove towards Contact Creek, I saw tons of wildlife crossing the highway, getting out of the fire area. Everything from porcupines to bears. You see a lot of wildlife up there, but usually they're crossing the highway

*both* ways. All of a sudden you're aware that you're seeing a *lot* of wildlife, and they're all heading north. They were all leaving. There was smoke in the air, but that had been there for weeks. So why now? *They* felt it as well. This was about three days before the fire blew. It made the hair on the back of my neck stand up.

Van Horlick was, in the most modest sense of the term, fighting the Eg fire at the time. It was an anomalous picture—the Eg by now covered tens of thousands of hectares but Van Horlick had only a seven-man crew working on a cluster of islands in the Liard near Contact Creek, attempting to prevent the fire's spread north into the Yukon. Though the size of the force assigned to the fire had varied over previous weeks, this was the only suppression action being taken on the Eg at the time.

*Tim Van Horlick, who fought the Cran fire before being reassigned to take what he describes as token action on the Eg. He remains a passionate critic of the Forest Service's handling of the Eg.* Lonny Miller photograph

On July 28 Van Horlick contacted a supervisor to say he had a sense that "big things were going to happen. Something was going to break loose." The air had been stagnant for so many days and there was "a funny feeling running through the whole countryside." Van Horlick was instructed to continue overseeing the nominal suppression effort near Contact Creek.

On July 29, the morning Brahniuk woke to his unsettling sensation, the week-old inversion began to dissipate. As the extensive thermal blanket lifted, inflowing air stoked the Eg like bellows on a forge. Flames built over a huge area. As hot air rose into a towering convective column, cooler air rushed in at lower levels to replace it, generating winds so powerful they flattened

*Firestorm near Fireside early on the evening of July 29. It looked like "a huge beach ball—a huge fireball—with fingers 600 to 800 feet high sticking up, just kind of rolling across the plain."* Ken Brahniuk photograph

whole sections of forest. From the higher land across which the fire was now blazing, gentle slopes ran down to the Liard and its floodplain, bearing magnificent stands of spruce. The Liard is a big river, but it proved no impediment to the fire's momentum. Van Horlick was airborne when the Eg took its running leap.

> Van Horlick: That big jump was something to behold. I'll never forget that. I was flying with a pilot by the name of Calvin Voth. We knew the fire was making a run and we were keeping an eye on it. It was spotting probably two miles ahead of itself, going downhill.
>
> We were over the Liard River in the helicopter and we could see the winds pick up as the fire came down the hill. It jumped across; from halfway down the slope it must have jumped three or four miles straight across the river onto the other hillside. The winds—fire winds—were tremendous, flattening the trees. It came down through some really heavy fuel and generated a tremendous amount of heat and convective winds.

The flames were really unimpressive—a few circles a hectare or so in size. The flames were all behind smoke, so it was unimpressive in terms of flame height. But that main column...and the wind speed...and seeing trees flattened...and getting caught in the helicopter with the winds...You're flying along straight, then you're virtually flying along on your side. The convection winds were at least 150 kilometres an hour when that fire ran.

Just before 3:30 on the afternoon of July 29, Brahniuk received a radio message from Van Horlick. My fire's up and running, the Eg boss said, and it's headed your way. Fast. Brahniuk immediately jumped into his helicopter to make his own assessment. Climbing through smoke, he was confronted by an electrifying sight: a 25-kilometre flame front bearing down on him and his crew.

*Fire activity on the Eg.* Douglas Cowell photograph, BC Forest Service collection

Brahniuk recalled, nearly two decades after the fact, the relief he felt when, from his helicopter seat, he contacted his second-in-command, Harry Hutchinson, "pushed all the buttons" to evacuate the Cran, and learned that everyone strung out along the fire's perimeter had received the word. A gravel pit along the Alaska Highway had been serving as the Cran's equipment base, and it was to that pit that the crew beat a hurried retreat.

As his firefighters fled the Cran, Brahniuk sent Hutchinson to Fireside to help prepare that community's evacuation. He also dispatched water trucks to block the highway as he realized that the narrow strip of gravel would in no way protect drivers caught in the Eg's path. Both emergency steps—evacuating a community and closing an international highway link—should have been approved through official channels, Brahniuk realized. The thing was, officialdom wasn't working—at least not in Brahniuk's isolated world. He couldn't contact the regional fire commissioner's office, and he didn't seem to be getting anywhere with a request for RCMP assistance in shutting down the Alaska Highway. His overwhelming impression—one that Van Horlick shared from his position at Lower Post—was of being left on his own to cope with a natural force of immense destructive potential.

In an uncannily similar replay of events on the 1971 Tee fire, Brahniuk found himself having to be concerned about a Northwest Tel microwave tower standing vulnerable on a nearby hilltop. Company representatives had visited Brahniuk on the Cran days earlier, telling him that, should fire endanger the facility, they would have to disconnect its power supply before it burned. With the Eg now moving in fast, Brahniuk flew to the site with a Northwest Tel technician, who disappeared into the tower. The fire advanced. The technician didn't reappear. Brahniuk's pilot—his name was Pierre—ran out of patience. "If that guy's not out here in one minute," Pierre told the fire boss, "I'm out of here." He didn't appear to be joking. About a minute later, his task completed, the technician sprinted for the helicopter. Brahniuk, holding the chopper door open, yelled for him to jump. The technician was still standing on the skid when the helicopter lifted off the hilltop. By some stroke of luck, the tower was not seriously damaged.

Half an hour later, with his crews withdrawn to Fireside, Brahniuk and Pierre were the sole people remaining on the Cran. The pair fuelled up their chopper, took off, then just as quickly set back down. The engine was malfunctioning. Draining the fuel system, Pierre found water contaminating what should have been pure Jet B. The fuel had been pumped from a sealed drum, so the pair knew that it wasn't necessarily an isolated case of contamination; there could be water in any of the remaining barrels.

The Eg was now about 15 kilometres away, but its presence was palpable. The sound, Brahniuk recalls, was "really eerie. It was a kind of whining, like something you'd hear in a horror movie. It's hard to describe, but I know it's something I'll never forget." The horror in the movie now playing was amplified by the fact that Brahniuk and Pierre had one means of escaping the oncoming Eg, and that was a helicopter operating on dubious fuel. Brahniuk commented that he didn't like the idea of flying in a machine that may have water in its tank. A practical man, as pilots tend to be, Pierre told Brahniuk, "We can take our chances in the air or get cooked on the ground."

*Ken Brahniuk took this photo of the approaching Eg fire as he and his crew hastily prepared to abandon their camp at Cranberry Rapids.* Ken Brahniuk photograph

*The Eg fire approaches Fireside.* Ken Brahniuk photograph

Airborne once again, Brahniuk had good reason to feel as mortal as he'd ever felt—a slingless David fleeing before a fire of biblical proportions. "It was probably the eeriest feeling of my life, being in the air in front of that thing," he later observed. It was obvious that, should Pierre be forced to put the machine back down, his "cooked on the ground" prediction would be realized. The fire front was now eight kilometres away, and Pierre commented that he could feel the chopper being drawn in by winds rushing to feed the Eg's voracious furnace. Both men could feel its heat through the aircraft's bubble.

Brahniuk jotted a brief fire log entry—an entry requiring nothing in the way of superlatives to communicate the drama of the moment: "1730 hrs. In helicopter. 600-ft. flames crossed river. Made one-mile run in one minute and totally engulfed CRAN Fire. Advancing on estimated 15-mile front. Ten miles from Fireside. Advised Hutchinson evacuate Fireside."

At Fireside, Hutchinson found a "pretty emotional" scene, with some people in tears at the prospect of abandoning their homes and possessions to the approaching fire. "They couldn't believe what we were telling them, what was coming at them," he later told photojournalist Doug Cowell. But when heavy smoke began rolling through town, disbelief was replaced by the survival instinct. People jumped into their vehicles and headed toward Coal River, 16 kilometres east of Fireside, near the confluence of the Coal and Liard rivers.

As the last person out of Fireside, Hutchinson verified that each building was empty. He taped doors shut after leaving each one, creating a seal that would be visibly broken should anyone attempt to re-enter. Several dogs had been left behind in the rush to evacuate, and Hutchinson began loading them into the back of his truck. One kept returning to its kennel, mystifying the forest officer until he looked inside and discovered puppies. With the fire now a roar in Hutchinson's ears, and with a truckload of forlorn dogs for company, he took to the Alaska Highway, playing the role of caboose in Fireside's evacuation train. As they passed their fire camp, the Cran crew stopped briefly at their tents, grabbed sleeping bags and food, and left the rest behind.

The Eg had now reached maximum ferocity. With the forest baked mid-summer dry, the terrain between the main front and the spot fires on its vanguard was bursting into flame. Pierre, it turned out, had successfully drained any water from the helicopter's fuel system and refilled with uncontaminated gas. The pair monitored the fire's awesome activity as they made their way toward Coal River. The air was filled with debris. Entire trees were inhaled into the maw of the convective column. The flame front, Brahniuk said, was like "a huge beach ball—a huge fireball—with fingers 600 to 800 feet high sticking up, just kind of rolling across the plain. It hopped over the river as if it wasn't there. Then it came right through town."

At Coal River Brahniuk quickly realized that everyone needed to put still more distance between themselves and the steamrolling Eg. Shortly there-after, a convoy of 175 people—25 residents and 150 bewildered tourists—pushed on toward Liard Hotsprings. Conditions on the highway, were "like driving through fog in a blizzard," Hutchinson reported.

Just before midnight Pierre, Brahniuk's pilot, announced that it had become too dark to fly and that he had no intention of losing his pricey machine to the Eg conflagration. As enterprising as he was determined, Pierre found two local men with a pickup who agreed to shepherd the chopper out of the danger area. Hovering just above the ground with landing lights on,

*A tree bursts into flames on the Eg fire.*
Douglas Cowell photograph, BC Forest Service collection

*Fire-ravaged Fireside Lodge with the Liard River in background. The lodge had been constructed during World War II by American servicemen building the Alaska Highway.* Douglas Cowell photograph, BC Forest Service collection

*One of the five homes destroyed when the Eg burned through the tiny hamlet of Fireside.* Douglas Cowell photograph, BC Forest Service collection

Pierre could just make out the pickup's tail lights through the drifts of smoke. He stayed just above and behind the cab, maintaining contact via two-way radio so that the men could tell him when he was coming toward rising ground or when they were about to pass through a canyon. "I talked to Pierre later," said Brahniuk. "He flew until he felt he was safe from the fire, then he landed. It took about an hour, and he said he was white-knuckled all the way—the toughest flight of his career."

The flight from the Eg was not much easier for the now baseless Cran firefighters, but they finally collapsed into sleeping bags on an abandoned airstrip near Liard Hotsprings for three or four hours' sleep. Brahniuk and Hutchinson, however, stayed up to formulate the next day's plans, their physical and mental functions nourished by a steady flow of caffeine and adrenaline. At 6:30 on the morning of July 30 Brahniuk returned to Coal River by helicopter to find that a shift in wind and a slight increase in humidity had prevented the Eg from overrunning the hamlet. Though still burning briskly for that early hour, the fire's behaviour was calm by comparison with that of the previous day. The good news at Fireside—astonishing news, perhaps, considering the force that had rolled through the community hours earlier— was that half of the town remained untouched by flames. The bad news was that the remainder, including the Fireside Inn, a garage, five houses and assorted outbuildings, were smoking ruins.

Flying over what only hours earlier had been the Cran fire, Brahniuk found a landscape scorched to a degree he had never previously encountered. Sloping up from the highway lay thousands of hectares that had been densely covered with small pine trees. Nothing of that forest remained. Everything organic had been incinerated by the Eg's tremendous heat, and the accompanying winds had blown away the resulting ash. "If I could envision flying over the moon or the Sahara Desert," Brahniuk commented of the sight, "that's what it reminded me of: grey, sandy-coloured dust. Not even stumps were left, just little black spikes where stumps had been."

Considering that the unburned portion of Fireside was no longer in immediate danger, Brahniuk set his sights on ensuring that Coal River didn't suffer a similar fate. Before doing that, however, he had to deal with the growing backlog of "antsy" northbound tourists and commercial truckers being held at the Liard Hotsprings roadblock. He remembers feeling vaguely presidential as he helicoptered in and was mobbed. Drivers did their aggressive

*When the Eg threatened the Alaska Highway, forest officer Ken Brahniuk was forced to make the unpopular decision of closing this vital northern route.*
Douglas Cowell photograph, BC Forest Service collection

best to convince him to reopen the road, but Brahniuk refused. "I didn't have time to argue with them. I insisted that we keep the highway closed. 'That's the way she is.' I finally got ahold of Dease Lake later and suggested that we try opening it up for a couple of hours. Vehicles went through, but it was scary. We were very lucky that nobody lost their lives."

As the Eg had crossed into a new jurisdiction overnight, the first call that the Fort Nelson Forest District got when it opened that morning came from Brahniuk. "Good morning," he told them, "you've got a new fire in your district." A few hours later he informed Fort Nelson of his plans for the day: establish fireguards around Coal River, burn off from those guards and, conditions permitting, light a defensive backfire. He then sent a 10-man crew to Coal River to pour water on buildings and surrounding timber in preparation for the burnoff. An aerial survey that evening, however, showed conditions too smoky to use an Aerial Ignition Device (AID), a helicopter-mounted machine that injects ethylene glycol into ping pong balls containing a small amount of potassium permanganate. The balls pour to the ground at a rate of seven per second before the chemical mixture ignites with an intense flame.

*Part of the extensive "moonscape" area where the intensity and internal winds of the fire burned and blew away all organic matter.* Tim Van Horlick photograph

Needing a fallback strategy, and feeling that any burn-off at all would improve the odds of saving Coal River, Brahniuk went to the community's sole gas station and filled several jerry cans with a blend of gasoline and diesel—gasoline for combustibility, diesel to control volatility. Perched in the back of a pickup, Brahniuk had himself driven along the highway's shoulder. As the truck rolled, Brahniuk poured the mixture onto roadside vegetation, stopping from time to time to light and lob out an incendiary flare. The resulting fires burned off much of the available fuels as they climbed the slopes above Coal River and likely contributed to preserving the town unscathed. As Brahniuk explained, "Instead of coming over top of us, the fire went around us." Though the Eg left Coal River intact, its presence that night remains etched in his memory. "I remember to this day sitting in Coal River at about two in the morning. You could read a newspaper from the flames around us, even though they were a mile or two away."

Brahniuk stayed on the Eg for two days after the big blow-up. He'd had little sleep during his five days on the Cran and none since waking to the spookily calm morning of July 29. On the evening of July 31, too exhausted to be of further use, he flew to the Dease Lake forestry office for a debriefing and to arrange replacements for himself and for Hutchinson, who was in a similar state.

To take charge of the Eg effort, the Forest Service selected an overhead team led by one of the province's top firefighters, Garnet Grimaldi. (The 1994 Garnet fire at Penticton was named in honour of Grimaldi, a resident of Penticton who died the day that fire ignited.) Grimaldi used the AID machine to light a series of burn-offs, thus preventing the Eg from regaining its former momentum. As control was established, firefighters moved into an extensive mop-up phase, strengthening fireguards and dousing hot spots, some of them detected with an aerial infrared scanner, but the fire's sweep over such a massive and remote area meant that much was left to burn out on its own. An infrared scan the following spring found areas of peat bog still burning.

When mapping was completed, the Eg's perimeter was found to encompass just over 182,722 hectares, making it the second largest forest fire in recorded BC history. (The fire boundary was later remapped by forest ecologist John Parminter, who concluded that it was only 126,000 hectares, though still the second largest fire that has burned entirely within this province.) Its smoke output was so prodigious that NASA researchers used it as a case study into the effects of nuclear winter. They estimated that as the Eg soot cloud passed over the eastern United States it lowered surface temperatures by four degrees Celsius.

Almost inevitably, smoke was still rising from the Fireside disaster area when burned-out residents and business owners accused the BC Forest Service of slipshod firefighting and fire management. A brief, futile struggle for financial compensation ensued. "One look at old burn patterns around there should show people that fire is a natural part of the ecosystem," the Forest Service's Paul Pashnik told *Maclean's* magazine. "Fireside itself sits in the middle of a pine stand that owes its existence to fire." But he later admitted, "We never thought the fire would cross the Liard."

Tim Van Horlick left the BC Forest Service in 1988. Now a private forestry consultant, he invented and markets a piece of machinery that prepares logged land for replanting. He remains a passionate critic of the way the Eg was managed, believing that the fire could have been successfully attacked prior to its explosive late July rampage. He describes the approach taken with it as "the waffle clause," which he interprets as meaning "'When it looks like it's threatening, we'll take some action on it,' which amounts to 'When it's impossible to do something, then we'll do something.'" In response, then-superintendent of fire control Jim Dunlop—who would become one of the most influential directors of BC's protection program— noted that the Eg "was one of those fires where we could have spent $3 million or $5 million putting it out and in the end found that we had saved only $1 million in resources."

It was not until Ken Brahniuk arrived home in the Queen Charlotte Islands that the magnitude of the experience and its attendant responsibilities hit him. When the impact sank in, it hit hard enough that he recalls having "a bit of a breakdown." Later, when someone mentioned his having flown by heli- copter to Liard Hotsprings for a shower and sandwich at some point during the fray, Brahniuk accepted the information as news. He had no memory of the incident.

One thing that stands out in Brahniuk's mind is the extent to which he and Hutchinson were left "basically on our own" to cope with the flaming beach ball that burned through Fireside, threatened Coal River and forced the closure of the Alaska Highway.

> Brahniuk: You just did what had to be done. To this day I don't remember getting a whole lot of support. Harry [Hutchinson] and I had this crew, and we had no radio communications other than radio-telephone to the Dease Lake Forest Service office. I had kept asking for help, saying that this fire was going out of control and we were evacuat- ing, but other than a couple of guys coming around, having a look and saying "Yeah, things are happening," and taking off again, we were by ourselves. I felt pretty isolated, like I was fighting a forest fire in Japan.

# — *Chicken Shit Happens* —

Searching for evidence that the final days of July 1982 were not unrelentingly grim, Ken Brahniuk can offer up only one anecdote, an event that could only pass as humorous when contrasted with the grimness of the days surrounding it. Like one of those party jokes where a message mutates as it passes from one teller to the next, the story worked its way through the Forest Service gossip mill, emerging as an account of how Ken Brahniuk "had the shit scared out of him" on the Eg fire. Brahniuk, however, tells a slightly different story.

> Just before the Cran was evacuated, our camp cook made chicken one night. It wasn't cooked well enough, but when Harry [Hutchinson] and I came into camp, we were so hungry we ate it anyways, and both of us got the runs really bad. At times, when we were out flying, we'd be getting our helicopter pilots to drop us down real fast—before they had a disaster on *their* hands.
>
> I remember one morning walking, kind of knock-kneed, past the Native crew to wash my clothes in the river, and them all laughing. They must have had stronger stomachs than us white guys. It never bothered them.
>
> That was about the only humour we had on the Eg fire.

# CHAPTER 10

# When Risky Business Goes Wrong

Fast-running fires, brittle snags, dislodged boulders, ponderous machinery. Dehydration. Precipitous terrain. Stinging insects. Fallen trees torpedoing down steep slopes. Aircraft operating in the narrow zone separating airspace from geology.

With the myriad risks inherent in wildland firefighting, it is perhaps surprising that the enterprise hasn't maimed or killed more people. It is difficult to determine how many firefighters, pilots and attack officers have died on the job, although the average is certainly less than one per year. BC has never experienced one of the mass-fatality tragedies that have burned themselves into American wildfire mythology. But we have come close more than once.

Training programs and minimum fitness standards have reduced the occupational risks. Today no one—not even in the direst of emergencies, as disgruntled loggers learned during the 1998 Salmon Arm crisis—is allowed to fight fire without having taken an introductory course offered by the Forest Service. But guidelines and training cannot eliminate the hazards lurking in wildland fires. Snags continue to burn. The terrain is no more forgiving. People find themselves in situations beyond their level of competence. In spite of sophisticated weather forecasts, Nomex suits and Workers Compensation regulations, all manner of risks lurk behind every new fire report.

From the Toronto *Globe*:

> Red Pass Junction, July 1, 1922. J. Bedford Edwards, a member of the British Columbia Forestry Department, was fighting a fire near here last Tuesday that had at one time threatened the beautiful Mount Robson Park, when, without a second's warning, a rockslide began ten feet from the brink of a 200-foot sheer cliff on the banks of the Fraser River, hurling him, amid an avalanche of huge rocks and debris, over the edge and down into the river, with both his legs broken.
>
> Joseph McCoig, an operator at Red Pass, happened to be in the vicinity, and, hearing the roar of the slide, rushed in that direction, arriving just in time to glimpse Edwards disappear into the deep waters of the river. He was powerless to go to the aid of the unfortunate man as there is no path or foothold from the top of the cliff. He watched with horror the struggles of the patrolman, who managed to battle at last to shallow water. McCoig ran for help and brought the section gang on the run who were working some distance away. The injured man had now been over 20 minutes in the water and it took ten men another 30 minutes to get him out and to the top of the bank.
>
> Dr. O'Hagan of McBride was wired to and he came in a special car, giving first aid to Edwards, who, in addition to having both legs broken, had cuts and bruises on head and body. Later in the evening he was transferred to Prince

George Hospital by Forest Ranger Lowry of McBride. The injured man is a veteran of the war, having been severely wounded while serving with the 18th Battalion in France.

A sad feature of the incident is that he was to have been married next week at Lucerne, a wire arriving two hours after his injury from his fiancèe, who had just arrived in Canada from South Wales. Should the diagnosis prove that there is a good chance of recovery, without loss of limb or permanent disablement, the young couple will be married at the Prince George Hospital in a few weeks' time.

Dave Procter grew up in the forestry town of Lumby and worked in a planer mill in high school before making a career with the BC Forest Service. He became known as a crack overhead-team administration boss, something he enjoyed a whole lot more than the stress of actually fighting fires.

> Procter: On the Black fire [in 1985, near Invermere] we had a guy struck by lightning. He had a hand tank pump—a metal tank on his back. He was climbing up the hill with it and he was tired, so he turned around and sat down. Lightning came down, struck the tank and, him being attached to it, struck him. I can't remember where it exited, but he was off work for a few days. His symptoms were that he couldn't see straight for a day or two—his vision was blurry—and his balls were all swoll up and sore.

Chuck Williams, a member of the Adams Lake Native band, was 14 when he got his first firefighting job, packing water on a fire being fought by members of the Civilian Conservation Corps near Whitefish, Montana. That was during the Great Depression when he was in the United States trying to find his father after a family breakup. He went on to become a cowboy on the Douglas Lake Ranch and later an expert faller.

> Williams: I got hurt in Golden at the Con fire. That was in '85. I had twisted my ankle on the Ram fire, so I had a week off. I went back up there with the wife to pick up my power

saw and stuff. I got up there and they needed firefighters. They were desperately in need of dangerous snag fallers. I came in with my power saw. They said, "You handle a power saw?"

I said, "Dangerous snag faller."

They said, "Helicopter's coming out here in 10 minutes. You'll be on that." I was in the helicopter before the wife even drove out of the parking lot.

I got hurt in a kind of a saddleback, a real bad place. It was my 22nd day on the Sullivan River. I had a snag come back on me. It was dinnertime on a Sunday when it happened. The first aid man told me what I had said after I got hurt: "I know well enough I should have gone to church this morning."

They took off my shoes and put me on this piece of plyboard. Somebody took a picture of me coming in under the helicopter at the end of a 200-foot cable. That thing was spinning around under the chopper. I came in on a wing and a prayer. Somebody told me later, "Over the drone of that danged helicopter we could hear what you were saying as you were coming in."

I said, "What was I saying?"

"Get me the fuck out of heeeere!"

When I got hurt there was a doctor, Doctor Hudson, the only one that knew the difference between a muscle and a nerve. He was in Toronto, so the compensation board sent me there for my operation. I was over there for a week. Then I was in the hospital in Kamloops for about six months. I told the doctor two times to take my danged arm off. But he said as long as I had a little movement in my fingers, that'll be good enough. I'm glad, really glad, he didn't cut it off.

They called it the Con fire. I told the foreman I didn't like that name. I said, "What the hell are they going to call us when the fire's over, ex-Cons?" He thought that was pretty cute. That was the last fire I was on. I enjoyed fighting fire. You never had the same thing the next day. It was always exciting.

In 1950 Glen Bertram was walking with a friend along a Penticton street when the pair were conscripted to fight a forest fire at Juliet Creek, along the Coquihalla River. Things came full circle when Bertram joined the Forest Service and did a certain amount of conscripting himself. His first job was as an assistant ranger in the East Kootenay region.

> Bertram: One of the toughest fires I had in Cranbrook was on Fisher Peak, a fairly large mountain. There was a little road up to an old mine. I sent a couple of guys up with power saws to cut out the windfalls, then I borrowed a quarter-ton army Jeep from the grazing department, something small enough to get up that road.
>
> I used to fight fire all day, then at suppertime I'd drive the Jeep down, get in my own truck, go into Cranbrook to pick up supplies—there were about 40, 42 guys on the fire. Then I'd drive back up to the fire about midnight. I forget how many days in a row I did that—for a couple of weeks at least. That put me in the hospital with hepatitis. I guess it was a combination of fatigue and living in dirt, camped out under a tree. Those were the kinds of things you went through day after day after day. I figured out that one year I slept out under a tree more days than I slept in a bed.

Bob Kenoras began fighting fires around Wenatchee, Washington, in the mid-1960s when government foresters would order out the sawmill crew he belonged to. He later worked as a faller on Nootka Island, off the west coast of Vancouver Island, but eventually returned to Salmon Arm, where he used his talents as a dangerous snag faller on the 1973 Glen Eden fire and other blazes.

> Kenoras: It can be dangerous work. You're the first person sent in before they send any men in to finish putting the fire out. You want it to be safe enough that if somebody's working there, no snag is going to fall on them. It could even be burning while you're falling it. If it's really burning bad, we just let it burn until it falls over. But if it's safe enough to fall it, then we get it down.

If you suspect a tree could be weak but it looks strong, then you take a bore cut out of it. You put your power saw straight in, and if you see smoke coming out of that cut you know you've got to fall it, because it's burning inside and it's going to keep on burning. If it's thin enough on the bottom—if you've got three-quarters of it burnt out and it's too dangerous to get near—what I usually did was go fall a big tree on it and bring it down that way. I don't know whether the Forestry frowns on that, but that's what I used to do. I got to the point with my falling that you could put a peg out anywhere around a tree—about three-quarters of the length of the tree away from the base—and I could bring the tree down on that peg.

I was working on a fire up around Malakwa when a snag came down on me. I'd just finished falling one snag, then I thought I could hear something. I thought somebody was coming behind me. I shut off my power saw and I was going to say, "What the heck are you doing here?" Because there's not supposed to be anybody close around a snag faller. I just started to turn around and look up the hill and here comes this snag.

I just stepped maybe a half-step back. My hard hat went flying. It took my wristwatch off. It hit my hand on the power saw handle, dented the handle down and threw me flying. I was knocked kind of silly for a while. Never broke no bones, but it scraped down my arm and all the blood vessels were standing up. I could hardly open my hand for a while. That's about the closest call I ever had. If I didn't look back like that I would have been six feet under.

As a heavy-equipment operator in the logging industry, Art Hart has fought fires for much of his life, mostly while living at Valemount. He worked for Slocan Forest Products until his retirement in 2001.

Hart: There was a guy killed on a fire near here around 1990, '91. He was filing his saw and a tree come down and hit him right over the head. He was leaning down beside a stump. The stump was a little bit higher than him. The tree come down over the stump, broke off, and nailed him over the head.

I was just a short distance away. There was guys standing right beside him, and they never saw the goddamn thing coming either. I gave him mouth-to-mouth and heart restoration. We kept him alive. We got him in a helicopter and into Prince George, but he never made it. That faller was from Quebec. A real nice little guy. A good worker.

Vern Hopkins was raised in the Okanagan Valley town of Westbank where, as a teenager, he participated in the BC Forest Service's junior forest ranger program. He was hired as an assistant ranger in the Kettle Valley in 1946 and spent his working life in government forestry, retiring in 1982.

Hopkins: We had about 75 soldiers go out on a fire in back of Westbank. We had a civilian packer who had two helpers, an Indian boy and myself. I think we were both 15. This would have been in 1943.

The horses were tied head to tail—one halter would be tied to the tail of the one in front. It was yellowjacket time and this log across the trail had a yellowjacket nest under it. The first horse across stirred them up. The second got stung. The third one decided not to go and he held back and pulled the hair out of the tail of the horse ahead. And I'm trying to hold this packhorse, which is lifting and rearing, and in the process of this I got kicked. When a horse is on very rough ground they add caulks—two little sharp spurs and a bar across the front of the shoe—to give them grip. This horse was shod in that way, and when he kicked me he punched quite a hole in my leg.

Meanwhile, the soldiers had gone on ahead, and the cook and all these supplies were going in with us. It was essential that they get in, so the packer was in a great rush. He made me comfortable beside the trail and continued on to camp. They sent someone back later, another young fellow a couple of years older than me, and he brought some grub, a can of beans and part of a slab of bacon. No cutlery or anything, so we whittled some sticks, toasted the bacon over a fire. We did pretty good. The following morning the packer stopped on his way back to town. He made rope stirrups for the pack saddle, folded a wool blanket across between the saddle horns—a pack saddle is pretty uncomfortable to sit on, just leather over a wooden frame—and that's how they took me out.

When he was 16, Mel Monteith lied about his age to get a suppression crew job in Kamloops. That was in 1947. Twelve years later he was handed his first ranger posting at Blue River in the North Thompson country. He retired in 1994 as manager of the Clearwater Forest District.

Monteith: Crowning fires are very spectacular. We had a very close call one time on a fire up the Raft River in 1971. We had a crew on it, an assistant ranger and one of the senior suppression crew men and about 20 other men. They were fighting this fire on top of a hill. And it blew up. It burned out part of the camp and it blew up where the guys were fighting the fire. Because they were experienced, the assistant ranger took his group out into the middle of the fire, onto a big rock knob that had already burned off.

I was flying towards the fire when it blew up. I said to the pilot, "It's going towards the fire lines! The crews are there! Hurry!" I flew around looking for fire crews, wondering how many men I'd killed in the fire.

The suppression crew man had taken his crew and headed out, well away from where the fire was running. The cook in the camp had gotten out of the way. And here was the assistant ranger and his crew on this rock out where the fire had already been, and they were fine too. Everyone was okay. But it's scary. You're losing your camp, and you know that your crew is fighting the top edge of the fire, which is something you sometimes have to do. But it would really upset you to lose a man in a fire.

Later in the 1970s we had a big fire at the north end of Wells Gray Park, at Hobson Lake. It had started on top and was coming down through very heavy timber. We wanted to stop it from burning right down to the lake, so we were just fighting it on that one edge. These were big trees, some cedars nearly a couple of metres in diameter, and we were building hand guard with power saws. We had a bunch of fallers cutting a big wide trail through this timber. This one young faller for some reason went down into a gully and didn't tell the other fallers he was going down there. And they dropped a tree on him. Broke his back.

Luckily we had radios at our campsite just down the hill, about two miles away. We had first aid people in there, but we were miles away from the nearest hospital in Clearwater. The problem was, how to get him out of there—heavy timber, steep gully. So we decided to take a chance. We got a helicopter with a long line and a basket stretcher and we lowered it down and lifted him out. You're not supposed to lift people in an open basket stretcher under a helicopter. That's a no-no. But it was either that or take an hour to carry him up to a heliport. We took the chance. The pilot said, "Gee, I don't want to do that." Jim Miller, the fire boss, said, "All right, I'll order you to do it. How's that?" The pilot said, "All right, you order me to do it." He was just excellent. Jim told me that when he set the stretcher down he barely bent the grass. He was that gentle.

Mac Morrow joined the BC Forest Service in 1958 and spent 10 years working around much of the province before he left to go commercial flying in Alaska. In 1974, while living in California, he was persuaded to return to BC as supervisor of the Kamloops air tanker base, a position he held until his retirement in 1997.

Morrow: The summer of 1974 there were seven air fatalities around Kamloops. People that you'd had breakfast with just didn't come home that night. So you'd go out and have a beer for them: "Let's have a beer for Keith." The next day you'd go back flying. Another night it would be, "Let's have a beer for Mac." That was a bad summer.

Pilots tend to be fatalists. What else could you do? The guys would feel bad. Sometimes they were so choked up they couldn't even talk, but we'd always go to the neighbourhood pub down at the corner. We'd have a couple of beers and go home. There was no use in dwelling on it. You didn't have time to sit around right then and let it get to you. At six o'clock the next morning you had to crawl back into your airplane. It got to you, all right. It was a kind of numbness.

You'd be curious. As a pilot you always wonder: What the hell happened? Did something happen to the airplane or did he make some kind of mistake? Normally it was a mistake. One time a pilot was coming up to Williams Lake from Abbotsford. He got into real bad weather in the Fraser Canyon and turned south. We had a search for five days, then they found him 50 feet from the top of Mount Stoyoma. Going up through clouds, he'd hit the top of this mountain. If he'd cleared the mountain he'd have been into flat country, Douglas Lake and that area. But he didn't punch his load out—that would have allowed him to climb faster. He came out of Abbotsford loaded with retardant, and he didn't punch his load out when he got into trouble. And this guy had just retired from the air force, search and rescue. He should have known better. Pilot error.

Jack Hogan was born in Nova Scotia and grew up in Fernie, a town destroyed by wildfire in 1908; his grandfather's family were among its victims. Hogan dropped out of Grade 13 to work in a coal mine, then sailed as a deckhand on a West Coast freighter before joining the BC Forest Service. He retired in 1987 and lives in Cranbrook.

Hogan: 1958 was a terrible year, lots of fires. We had one we called the Paul fire, north of Kootenay National Park, before you get to the Ice River. It was started by lightning. A fairly wet area, so you had big, big spruce.

We got some fireguards in, then two things happened to defeat us. First, we had an inversion, and you've got to live through one to know what that's like. We were sleeping way up at 6,500 feet elevation, and we were warm with one blanket halfway up our bodies at night. The temperature was about 70 degrees. And they pretty near froze down in the meadow where the base camp was, eight miles away, down along the Kootenay River. So there was cold air above, cold air below, and hot air right where the fire was. The thing glowed all night. It was like a well-lit city—flames here, there, everywhere. We had to keep the heck away from it, 'cause there was stuff falling everywhere.

Then we got another frontal system, and all the terrific winds that happen in a lightning storm. With those two things combining, the fire jumped a creek where we had it corralled. It crossed the other slope and went right up to the top. The next creek was where our camp was, just hung there on a hillside. You had to dig a hole to make a flat place to sleep. People bivouacked where they could.

We were trying to save what guards we had at the base of the fire and up into this canyon. One thing you need to learn about fires is to listen to the old people. I didn't in this case. One old-timer told me, "Don't go into the canyon you're in. Go into the next one over." Well, there was a whole bunch of rocks to get over to get to the next ridge. The canyon we were in was almost void of fuel but we had to make ladders to get up it—fall a tree and cut notches in it, then three or four guys would manoeuvre it into position and you'd climb up. So we had this staircase going up this canyon to get up on top. Then this freakish storm came along and blew the fire up this other slope, so we had fire on two sides.

Hopkins [forest ranger Vern Hopkins] said to me, "We'd better take you out for a rest." He pulled me off the fire for the day. I told him, "You'd better get some more people here. We've got twice the acreage burning than we did when we started on this thing."

Steamboat Lumber had a crackerjack bunch that was working on a fire up Forester Creek, and they had that one pretty well handled, so they were going to come in and help us. This crew came in, and Eddy Paul was with that bunch—the extra troops. He was a chainsaw man, a faller-bucker, a Native Indian. They were working on this canyon thing and we had all kinds of snags to deal with, some of them burning.

This young Paul felled a tree, a pretty good-sized spruce. It went over and bent another tree over as it fell, and when the bent tree stood back up, it broke a dead balsam off. So this piece of balsam came over. Meantime, Paul had just turned to his helper when his helper hollered for him to watch out. But instead of just falling, he looked up to protect himself—to do something—and this thing came down and hit him diagonally across the neck and the back. It broke his neck, and in about four minutes he was dead.

At the request of forest ranger Vern Hopkins, the name Paul Creek was officially adopted for a previously unnamed watercourse flowing into the Kootenay River northeast of Spillimacheen. Edward Paul was an East Kootenay Shuswap Native and a father of four children when he died fighting what was officially known as the Pin fire, named for Pinnacle Creek.

Hogan: We had a fire in 1960 in Kindersley Creek, which is up towards Edgewater, in the Rockies. This was another steep fire. We pretty well had it corralled. We had favourable weather for a few days and we wanted to extinguish it, so we put in a relay pumping system, pumping from the creek up to a landing on the edge of the fire. From there we split the hose and got water wherever we could, all around the fire.

This man Norman Johnson was a foreman of the afternoon shift—we were wearing everybody out so we had put an afternoon shift on. He'd just been down to the base pump. They'd had a little trouble and he'd helped the guy down there get it going. He had just climbed back up to the pumping station on the fireguard. There was another man coming down to see when they were going to start the pumps, and he hollered, "Look out!"

The roots had burned off one of these danged spruce or balsam trees. This doggone thing came like a whisper from the standing timber inside the fireguard. There was a kind of dip where the pumping station was. Well, Johnson straightened up when the guy yelled, and he got full whack. If he'd fallen down beside the pump, he might have just been injured—punctured by a limb or something. But it whacked him across the head, neck and shoulders.

[Assistant ranger] Bob Sahlstrom was in charge of this fire. He called me and said, "There's been an injury."

I said, "Holy cats! What have you done?"

"I've phoned Doctor Duthie and he said we could take his station wagon." We didn't have ambulances standing by at our beck and call in those days. [According to former Invermere ranger Vern Hopkins an ambulance had been assigned to the fire but the day crew had driven it away to use as their transportation. The resulting delay sparked considerable criticism of Hopkins and the Forest Service.]

I said, "Doc Duthie won't make it up there. He'll only get to one of the landings near the fire, then we'll have to work from there." So we got a stretcher, got Johnson down to a four-wheel drive. A couple of guys sat with him and cradled the stretcher on the way down. This was rough going, a bulldozer trail. We got him down to Doc Duthie's car, and he said, "Bob, you'll have to hold onto the stretcher. Jack, you drive. I'll stay in the back with the patient."

Johnson said, "Gee, Doc, you'll have to give me a couple of aspirins. I've got a hell of a headache." Christ, he had a brain concussion. We nursed him down, gunned him in to the hospital in Invermere. He spent some time there, then they lined up an ambulance and crew and decided to send him through to Calgary.

We didn't learn until later that Johnson died in the ambulance at Eisenhower Junction, where Highway 93 meets the Trans-Canada Highway.

Bob Sahlstrom participated in Norman Johnson's evacuation to hospital, and later returned to inspect the accident site. The tree that killed Johnson was a large spruce that looked healthy from the outside and therefore had not been brought down by fallers as they removed snags that posed a potential danger to firefighters. The tree had a rotten core and had burned from the inside out. After falling, the tree came to rest about two metres above the natural depression where Johnson had been operating his pump. As the tree struck the ground it would have flexed and struck him; his hard hat would have been insufficient protection against the tremendous impact. Sahlstrom remembers that breaking the news of this accident to Johnson's wife was "the most difficult thing I've had to do in my life." He also remembers that during July, the month that this event took place, he recorded 397 hours on the job, putting in many consecutive days that began at 4:00 a.m. and continued until 11:00 p.m. "Back then," he said, "fighting fire was war." Sahlstrom left the Forest Service at the end of the 1960 fire season to attend the University of British Columbia. He made a career teaching special-needs high-school students in Castlegar where, since his retirement, he has co-managed a 600-hectare wood-lot operation.

Based on the precedent established with the death and subsequent honouring of Eddy Paul, ranger Vern Hopkins applied for a second geographic memorial. The name Mount Johnson already existed, so in 1965 an unnamed peak in the Brisco Range, the divide between the upper Kootenay and Columbia rivers, became Mount Norman.

# CHAPTER 11

# Trout à la Suisse

"If a particular weather event has never before occurred, at least in recorded history, it simply would not be expected."

—from federal weather technician Daryl Brown's report on "the extreme weather conditions" of May 27–31, 1983

For several days prior to May 27, 1983, the weather on BC's Interior Plateau had been cool and showery, and more of the same was forecast. As a result, Rod DeBoice, a forest officer based in Houston, was questioning his choice of time-off activities. "My wife and I and some friends from Alberta were at Takysie Lake, a great rainbow trout fishing lake. I remember it being bitterly cold that Friday afternoon. We had down vests and gloves and toques on and were joking about how crazy we were to be out fishing in this weather."

At some point that day a sudden weather shift sent a flow of unusually warm and dry air streaming into central BC from the southwestern United States. Temperatures soared into the mid-30s, about 20 degrees above the seasonal norm. Humidities dropped into the low teens. Buoyed by shirt-sleeve conditions, DeBoice and his party trailered their boat and headed south to sample the char fishing at Ootsa Lake.

# Trout à la Suisse (Swiss Fire, 1983)

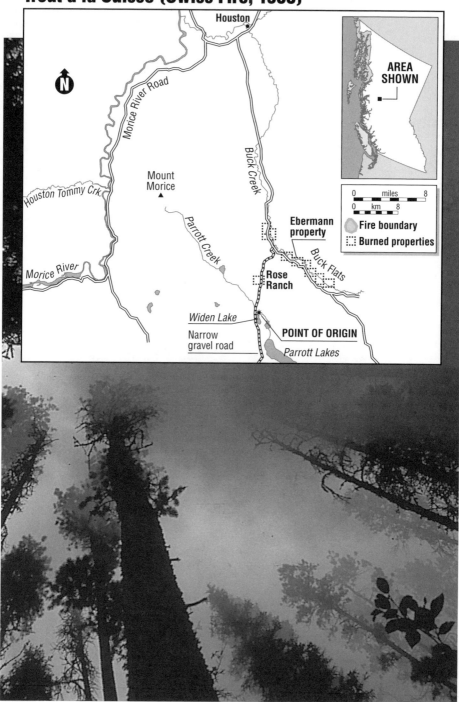

Not far from the DeBoice party, Christian Kaiser and Werner Hugentobler were also making the most of the sudden and unseasonal shift. The 23-year-old Swiss backpackers had pitched their tent at an informal campsite on Widen Lake, accessible by logging road, approximately 30 kilometres south of Houston. It was in regular use by Bulkley Valley locals, by tourists seeking solitude beyond the beaten path and by area Natives who netted and smoked fish there. The Swiss men had been guided to the spot by a local Native man who had shown them how to smoke their fish in a wood-framed, plastic-covered smokehouse that had already been built near the lake.

*Trout from the smokehouse where the Swiss fire began.* BC Forest Service collection

Kaiser and Hugentobler were alone at Widen Lake on Sunday, May 29, two days after the weather anomaly anchored itself over the Interior Plateau. While their trout smoked that day, the young men settled down to read over the brow of a nearby rise. A wind sprang up. The smokehouse fire, which had until then been confined to a hole in the ground, ignited the makeshift structure. Arnold Amonson, manager of the Morice Forest District, would later describe in an affidavit what the men told him as they sat in his office early the next morning: "They looked over to the smokehouse and saw flames in the trees. They tried to extinguish the fire using a 10-litre pail, but the fire was too far up in the trees for them to reach."

While the backpackers watched flames vault up a large pine tree and from there into a stand of young spruce, Bruce Noble was in the air over Nanika Lake, 80 kilometres to the southwest. A Forest Service air-attack officer, Noble was directing a bomber group consisting of two A-26s and a Firecat as they dropped retardant on a fire dubbed the Nan. Being closer to the Coast Mountains in wetter country, Noble noted, the Nan turned out to be a relatively easy piece of work.

Noble: The A-26s are quicker aircraft, so the Firecat was trailing them. I actioned the Nan with the A-26s, and just as

**171**

I was finishing up with them, I got a call from the pilot of the Firecat saying he had spotted smoke up towards Parrott Lake [Parrott Lake and Widen Lake are situated near each other along Parrott Creek]. I said, "Okay, fine. We've finished with this one. It's not a problem." And we diverted to the Swiss fire.

Rick Fiddis had watched Noble's A-26s lay down their retardant lines at Nanika Lake. An inventory resource officer with the Morice Forest District, Fiddis had been overseeing a ground crew on the Nan when over his radio he heard the discussion about smoke near Parrott Lake and Noble's subsequent decision to investigate. With the Nan in mop-up stage, Fiddis had his helicopter pilot follow Noble toward the new blaze.

Noble, being flown in a Piper Aerostar birddog plane by Conair Aviation pilot Andy Walsh, arrived from Nanika Lake when the upstart fire had been burning for about ten minutes. He estimated its size at six hectares and noted that it was moving fast and spotting well ahead of itself.

Noble: We were still experiencing a bit of the overwinter drought. That area hadn't seen green-up yet. The deciduous hadn't flushed, the new grass hadn't come in. The fine fuels on the ground were cured and available, so it made for very volatile fire behaviour. There wasn't a lot of wind at that time but the fire had enough intensity that it was starting to create its own wind.

Noble had Firecat pilot Clyde Blyth drop his retardant load near the recreation area where Kaiser and Hugentobler had been camped. By now the remains of their packs, fishing equipment, boots and some clothes were smouldering alongside the lake. The pair had moved into a swampy area, as far out of harm's way as they could immediately get, and from there they watched the forest downwind of them explode.

Meanwhile, the A-26s in Noble's air group had flown to a tanker base at Smithers for retardant reloads; Blyth's Firecat followed them. Noble briefly surveyed the activity below then, being low on fuel, directed Walsh to Smithers as well.

On-scene minutes later, Fiddis looked down from his helicopter at a blaze that was, he reported, "totally uncontrollable" and spreading fast to the northeast. He radioed Arnold Amonson, requesting that ten Cats and 100 firefighters be organized for first light the next morning. Amonson, in turn, put in a call to Northwood Pulp and Paper, one of the area's two major timber producers and the company on whose tenure area the fire was burning. He told the company to mobilize heavy equipment and provide a person to oversee their part of the suppression operation. Northwood's fire specialist was Dan Metcalf, a former Forest Service employee who had worked under Amonson in northern BC before moving to private industry. Metcalf was setting up a sprinkler on the lawn of his corner lot house in Houston when Amonson drove up, stopped in the intersection and rolled down his car window. "Don't go away," Amonson said. "We're going to be calling you." Metcalf looked to the south, saw what he described as "a Jesus-big white cloud," and knew that his weekend off was over.

*The Swiss fire ignited before spring's lush growth had arrived in the Houston area, a factor that helped set the stage for a destructive rampage.* BC Forest Service collection

While Amonson pulled together firefighting resources, Noble's air group took off from Smithers and got a good look at what Noble described as "basically a holocaust. It was up in the trees, it was spotting, and it had a column up around 10,000 or 12,000 feet." Less than an hour had passed since Kaiser and Hugentobler had looked up to notice something unusual going on at the smokehouse. Noble concurred with Fiddis that the fire was, for the time being, beyond resources, much as the Eg fire had so quickly become the previous year. And as a host of new fires were being reported throughout the BC Interior, Noble and his trio of air tankers were dispatched to fight them.

Fiddis had his helicopter pilot land at the fire's point of origin. There he found the two young men still standing beside Widen Lake. After gathering a few particulars, Fiddis gave the pair the best advice he could: get in your canoe, paddle down Parrott Creek to Parrott Lake and keep going to the end. Before pushing off, Kaiser and Hugentobler had a question for Fiddis, one that must have occurred to them as they stood, bucket in hand, watching their smokehouse fire roar into the surrounding forest.

"They wanted to know what was going to happen to them and if they would go to jail," Fiddis reported. "I replied that I did not know and to just get moving in their canoe and that I would arrange to pick them up later."

Back in the air, Fiddis saw the Swiss blaze marching steadily northeast toward several isolated farmsteads strung out along Buck Flats Creek. Separating the fire from Buck Flats was a line of ridges running generally north-south. Located on one of those ridges was the Rose ranch.

From Missouri via Washington's Vachon Island, the Rose family—Lee, Dorothy and three of what would ultimately be five sons—had moved to Canada in 1968. Their property, privately owned acreage in the midst of crown land, had been homesteaded decades earlier by Emil Widen, a teamster attracted by the natural meadow that provided feed and a resting place for the horses he used to freight ore to Houston from a nearby mine. For $5,000, the Rose family got a quarter-section of spectacular isolation and five dollars change. They improved the homestead as money and time allowed. To cover expenses, Lee worked as a high-school math teacher in Houston, 30 dirt-and-gravel kilometres distant.

Lee was doing chores on the afternoon of May 29 when he heard first one, then a second plane overhead. Their trajectory led his eye to the vicinity of

*The Swiss fire.* BC Forest Service collection

Widen Lake, out of sight just below the Rose property. He could see smoke rising near the lake, and thought he noticed one of the planes drop something near that smoke. Curious, he and two sons drove down the rough track leading from their farm to the lake. Partway there, he hit the brakes. A wall of fire was advancing toward them, the flames extending well above tree height. The two-rut road forced him to do a fancy cut-and-fill about-face, his calm being quickly eroded by his son Richard's exhortation, "Oh my God, Dad, don't get stuck!"

A late afternoon reconnaissance showed the Swiss fire burning in second-growth conifers and scattered stands of aspen. Its behaviour had calmed somewhat, causing forest district officials to set the following day's objective at containing the fire at its existing size.

That optimistic assessment had barely been completed when a freshening easterly wind blew the fire into the old-growth timber that dominated the largely unlogged area. The forest district's fire protection specialist called for air tankers to lay down a 1,300-metre retardant strip in the fire's path. He was told by superiors that the necessary aircraft were committed elsewhere. Uncontained, the fire raged west over Mount Morice and toward the Morice River. Rod DeBoice, having been pulled away from his weekend of fishing to help lay the logistical groundwork for a major fire campaign, drove out the Morice River road that night to see first-hand what the excitement was about.

> DeBoice: The fire was burning over Morice Mountain when I went out. It was quite an experience to see the glow of a fire front a few kilometres long coming over the mountain. The sap hadn't started running in the trees yet so they were popcorn dry. The grasses and other fuels were tinder dry from being freeze-dried over the winter, so conditions were explosive. Cottonwood trees were exploding from the radiant heat and showering embers for hundreds of feet around them.
>
> The thing that stands out in my memory is the amount of wildlife that was scattered all over that Morice River road. It had come off Morice Mountain ahead of the fire: moose, deer, bears, the odd wolf, anything you could think of.

Some of them were within a hundred feet of each other, standing between the Morice River road and the river. I guess they were deciding whether they were going to try to cross the river, knowing they couldn't go back up the hill.

The river was rising by the hour as summer-like temperatures triggered rapid snowmelt throughout its watershed. It quickly became apparent that the Morice was more of an impediment to wildlife than to the flames, as with little pause, the fire spotted across the current. Nearly 7,000 hectares of forest were now burning, including a sizable area of roadless conifer stands stretching away from the river's western bank. As there were no bridges across the river in that area, there was no prospect of immediate containment on that side. East-flank control was still considered possible, however, and by midnight, as an overhead team

*Lee Rose and grandson Alex Rose in 1999. "The fire was a struggle that not only united Dorothy and I," Lee said, "it brought us closer to the whole town."* Keith Keller photograph

assembled in nearby Smithers, a dozen pieces of heavy equipment were building fireguards between the fire and Buck Flats, just below the Rose ranch.

Returning to the farm after his encounter with the flames above Widen Lake, Lee Rose alerted his family, including his mother who lived in a separate house, to the danger developing below them. Smoke had now risen high enough that some neighbours along Buck Flats Creek assumed the Rose house was on fire and drove up to see how they could help.

Lee and Dorothy Rose spent much of that night watching the flame front advance. Its behaviour is imprinted in Lee's memory. "It was burning so furiously that we saw whole trees ripped out of the ground, roots and

*Bucketing retardant on a hot spot.* BC Forest Service collection

*Charred trees deposited in a cyclonic pattern vividly illustrate the powerful winds that develop in intense firestorms such as those seen during the Swiss fire.*
BC Forest Service collection

all, tumbled hundreds of feet into the air and then suddenly disintegrate into a cloud of smoke." The fire's direction gave him some hope, however. "I thought the fire would miss us. It looked very clearly that the fire front had gone by to the west, and that we were out of danger."

Lee's optimistic outlook was bolstered by the fact that the slope below the ranch consisted of pasture in its first flush of spring growth. Forest encroached on the homestead only on its northern side, and Lee, studying the fire to the southwest, couldn't imagine flames outflanking the verdant island that otherwise surrounded his home. His confidence was such that he rejected neighbours' offers to haul to safety the family's furniture and other portable possessions.

Throughout the fire's initial stage, while Lee, the man of the house, was being consulted by a stream of neighbours and Forest Service personnel, only one person thought to solicit Dorothy Rose's opinion on any point at all. "I'm still ticked off about that," she admitted, the soft edges of her vestigial Missouri accent hardening. "But," she noted, "one smart man thought to

*A section boss gives orders to a crew boss as firefighters attempt to outflank the Swiss fire.* BC Forest Service collection

come to *me."* Dorothy told that smart man that he should pack off her kitchen table and her rocking chair, just in case.

Just in case, as it turned out, was just around the corner. That morning the Forest Service had received a forecast calling for "extreme fire behaviour" to be expected later that day. Indications of that potential were easy to find. At five o'clock on the morning of May 30, when Rick Fiddis relieved Dan Metcalf as supervisor of fireguard construction along the crucial east flank, he found that he could drop a lighted match to the ground and see the resulting flames climb almost instantly to the top of nearby trees. The question, therefore, was not whether there would be spectacular fire activity once the day heated up but where flames would go once they started running.

For as long as people have been put in charge of fighting wildland fires they have had to deal with the question, At what point do I abandon the suppression effort in favour of protecting firefighters' lives? Fiddis had to face that question later in the morning when he elected to pull men and heavy equipment back to the Rose ranch. That was when a member of the overhead team taking over the fire arrived and questioned Fiddis's decision.

What ensued was what Fiddis recalled as "quite a discussion," during which he argued that Cats were operating too close to a volatile fire front with no good escape routes. "I saw what it did yesterday," he told his counterpart, "and it's warmer and drier and windier coming up." His final message to the O-team member was this: "When you take over the fire and I leave, follow your plan. Until then we're following mine." Fiddis had formulated his own fire behaviour model for the day, a prediction most easily understood in layman's terms—"It was going to get up and fuck off on us." He had Cats and crews pull back first to the Rose ranch, then to Buck Flats, where they would attempt a new stand among the series of small farms located in that valley bottom. In retrospect, Fiddis recalled his insistence to withdraw as "one of the better decisions I've made in my life." He added, "I still feel good about that one. We weren't going to have any effect on the fire anyways. It had a mind of its own; it was going to do what it was going to do. It would have killed people if we hadn't gotten them out of the way."

While his lieutenants developed fire strategy that morning, Arnold Amonson was working against the clock to keep Christian Kaiser and Werner Hugentobler in town. Amonson told the pair that if the facts were as they described, they wouldn't face legal repercussions. "If a building burns down and the fire escapes into the forest, that's not against the law," he told them. "It's an accident."

Still, an investigation was warranted, to which the Swiss men were key. But, told that they were not being held, the backpackers said they would be leaving immediately for Alaska. They then asked Amonson whether, for insurance purposes, he would provide them with a letter explaining that their camping equipment had been destroyed in a fire. Maybe later, Amonson replied. But later wouldn't happen. Just before noon, Amonson found the backpackers at the Houston train station. Would they reconsider? he asked. Would they stay long enough for a proper investigation to be completed? No, they replied. In a last-ditch effort, Amonson offered the pair work fighting the fire. One of them told Amonson that he was asthmatic and couldn't stand smoke. The westbound train departed for Prince Rupert with Kaiser and Hugentobler aboard. (The men were never convicted with having caused the fire, a fact that reportedly elicited a diplomatic complaint from the Swiss embassy over the name that the Forest Service had given it.)

# — *The Security Lady* —

Helen Lokken was settled into an outdoor family gathering near Topley, between Houston and Burns Lake when, glancing to the southwest, she noticed smoke smudging the otherwise clear spring sky. She turned on a radiophone and quickly confirmed her suspicion—a major forest fire had broken out and a major effort was being organized to fight it. Lokken predicted that her afternoon at the lake was about to be cut short; three hours later she was working the Swiss fire.

A resident of Houston, Lokken worked on an as-needed basis as a forest fire security person, a job necessitated by the encompassing fact that, as she put it, "There's lots of things that go on on fires." Working fire security demands multiple skills, and Lokken was a natural—even-handed, hard-nosed, with a slow fuse and a ready sense of humour. Job responsibilities include holding rubberneckers at bay, intercepting camp-bound booze, preventing thefts. Lokken once had a fellow security person flip out when the passengers in a car he'd stopped refused to get out of their vehicle. His solution—pull a gun on them. "I damn near died," Lokken recalled. "I said, 'What are you *doing*?' He said, 'They get out of the car *now*'."

Lokken walked calmly to her own vehicle and radioed for help, which arrived before any shots were fired. "There is a lot of stress," she acknowledged. "You're worried all the time. People are mad at you. They're not realizing that your job and the firefighters' job out in the bush—they're the same thing. You're both fighting fire."

> Lokken: You know, you learn quickly who's going to do what. There's a lot of stealing on big fires. There's lots of things left out in the bush that are dug under; firefighters will bury stuff and go back for it later. The food trucks go up and you have to make sure that nobody's bothering them. And when the food trucks go back, you've got to check to make sure there's no food in the truck 'cause people are stealing food that way, too. The

truck is a big truck that goes in to SuperValu and loads up all the groceries they need for the camp for a week. And they may not drop everything off. Or the cook will steal stuff when he goes out of camp.

I found a lot of booze that they were trying to get into the Swiss fire camp. You can't blame the firefighters. I don't know why they can't set up something for the guys, at least let them have one or two beers after a 16-hour day. Because *Forestry* was drinking. And I reported that. I got mad. I said, "That's crap. The forest officers are sitting in there having a beer. Our men are out working their asses off." I looked at those men coming in from the fire, and they were *tired*. And they're making seven bucks an hour. It really made me mad. So I called the Forestry on it. I didn't care. I went in to Arnold Amonson, and Arnold straightened it up.

Some of the guys are very good with you. Some aren't. One night on the Swiss fire I was working at 17 Kilometre, way out on a gravel road in the bush. And it was dark. God, it was dark out there. Forestry had forgotten to bring me a reflective vest, and I didn't have a sign warning people to stop. So I'm standing on the road, trying to stop vehicles. And here come a bunch of guys back from town, drunk, and they damn near hit me. I dived into the ditch.

They stopped and backed up. They were shaking.

I said, "It's okay. Get out of the car."

They got out of the car kind of wobbly. "We bin drinkin'."

"I noticed."

They said, "We're going right to camp. We're building you a sign. We'll be back in 20 minutes."

And they did. They come back with a sign. They come back with flashlights. They come back with a reflective vest.

They were looking after the security lady.

Helen Lokken died in 2001. For most of the time since the Swiss fire she worked as a custodian at Houston Secondary School.

As Monday morning wore on, Dorothy Rose had son Richard drive her to Houston for sandwich makings to feed the growing numbers of fire-fighters and concerned neighbours gathering at the ranch. Returning from town, the pair found a roadblock erected across Buck Flats Road and a security person who was unsympathetic to Dorothy's demand that she be allowed to pass. A soft-spoken woman with an iron streak of resolve, Dorothy told Richard to ram the roadblock. Richard put the truck in gear. The roadblock opened.

Dorothy never did get to make those sandwiches. Shortly after arriving home, a Forest Service officer informed the Roses that they had five minutes to grab what they could and evacuate. A flame front was headed their way in high gear. Lee balked at the suggestion, saying he intended to stay and protect his home. The forest officer told Lee he would ensure that the house would be protected. Lee accepted the promise and joined his family. Among the few things they were able to gather before fleeing was Dorothy's parrot, Mate. Apparently sensing the urgency of the moment, Mate perched in the back seat of the car and put his limited vocabulary to good use, squawking "Help! Help!" all the way to Houston.

By early afternoon of May 30, the Swiss blaze was spreading on all perimeters and was judged "too hot for control forces." At one point, air tankers were dispatched in an attempt to hold the flaring east flank, but the planes departed without dropping their loads after the air attack officer leading them saw what he described simply as "the magnitude of the situation."

It was at about this time that the overhead team adjusted its day's objective to meet the rapidly developing crisis: save life and property and keep fire out of Buck Flats. The Provincial Emergency Program was engaged and Buck Flats residents were informed that their homes stood in the path of a potent and fast-approaching danger. That warning set into motion a northbound convoy of vehicles loaded with household possessions. Some towed trailers containing whatever cattle and horses could be rounded up. Other farm animals were left to fend for themselves.

In a last-ditch attempt to keep flames from running over the intervening ridges and into Buck Flats, backfires were lit along bush roads and fireguards east of the main fire. It was a desperate attempt to deprive the front of fuel, and a strategic decision that would later come to haunt the Ministry of Forests.

Recognizing that backfires would probably not prevent the fire's invasion of Buck Flats, the O-team soon revised its primary objective to the protection of human life; property losses would have to be accepted. Rick Fiddis's decision of that morning was now being adopted on a larger scale. All firefighters were ordered to pull out of Buck Flats. Those in a position to evacuate directly to Houston did so, while those cut off by the fire's advance took a circuitous route through the area's extensive network of gravel backroads.

In the final hours of Monday, May 30, as the Rose family and the Buck Flats exiles settled uneasily into a vacant townhouse complex in Houston, fire managers gathered to

*Dorothy Rose making butter, with a hillside scarred by the Swiss fire visible in the background.* Photograph courtesy the Rose family

assess losses and consider new lines of defence. Their status report noted tersely, "Fire size has spread to approximately 16,000 hectares. All lines lost. Fifty families have been evacuated from Buck Flats area. North flank of fire approximately 8 kilometres long and approximately 12 kilometres south of Houston. Fire moving north at approximately half a kilometre per hour."

Houston's 4,000 residents had become increasingly aware of the fire as Monday wore on, their sense of urgency heightened by the arrival of the Buck Flats evacuation convoy. Lineups formed at gas stations as people prepared for a quick getaway. Municipal authorities issued a warning that water should be conserved for firefighting, and a loudspeaker truck drove through the streets, warning Houstonians that, though they were not yet in "immediate danger," they should stay tuned to the local radio station for breaking news. With tinder-dry forest stretching from the steadily approaching fire all the way to their backyards, residents along the town's southern edge prepared to join the Buck Flats flight. By midnight Cats were bulldozing a "contingency

*Men working from a tank-equipped skidder douse embers.*
BC Forest Service collection

guard" 30 to 50 metres wide between Houston and the advancing front. The guard was roughly the same width as the Morice River—which flames had so effortlessly crossed the day before—a fact that may have been in the minds of some of the people tracking the fire's progress.

And then it was over. In a manner of speaking. Sometime during the night of May 30, the unprecedented weather anomaly that had established such spectacular burning conditions over the Interior Plateau four days earlier dissipated as abruptly as it had formed. Whereas firefighters had been overwhelmed by scorching heat and Saharan humidity, that trend suddenly reversed, with daily high temperatures falling and humidity rising. Three days later, fire crews worked in the rain, the mercury barely reaching double digits. A few days after that Northwood Pulp and Paper's Dan Metcalf and his helicopter pilot flew in snow as Metcalf "paper-trailed"—tossed out rolls of cash-register paper which would snag in trees and unfurl to the ground, pinpointing hot spots for mop-up crews.

On May 30, however, the Swiss was down but clearly not out. Many areas within the fire's perimeter continued to burn, and no one could be certain that hot, windy conditions wouldn't recur. Knowing what an escaped smokehouse fire had developed into, firefighters could barely imagine what would happen if similar conditions were re-established around the extensive blaze.

But if anything positive had come from the initial 36 hours of intense fire activity, it had been the massive mobilization of firefighters, equipment and logistical support. Getting fire-line labour hadn't been a problem. The province was in recession and unemployed people had lined up at the vacant Houston furniture store that Rod DeBoice had transformed overnight into a fire command centre. By June 2 the Swiss fire force consisted of 500 people backed by 48 Cats, 30 skidders and a dozen helicopters. With the fire suddenly calm enough to approach on most flanks, all these resources were thrown into gaining control.

As activity gradually shifted from control to mop-up, Forest Service officials began focussing on rehabilitation and salvage. Rehabilitation consisted primarily of establishing erosion control devices known as water bars on the miles of fireguards bulldozed to bare soil, then stabilizing the guards by seeding them to grass.

Though forests are commonly reported to be "destroyed" by wildfire, in most cases much wood continues to be harvestable for a year or two before drying splits and twists it beyond the point of economic recovery. Thus, an intense salvage operation reduced the volume of timber lost to the Swiss fire. Still, just over a million cubic metres of timber was reported as being unsalvageable, nearly as much wood as is processed annually at Houston's massive Canfor sawmill-planer facility. (Canfor purchased Northwood Pulp and Paper in 2001.)

Northwood, having tenure in 75 percent of the burned area, logged and milled much of the burned Swiss timber. In addition, a host of independent operators set up small portable mills in the fire area, creating something of a gold rush atmosphere. However, a combination of ash, inexperience and depressed lumber prices made the venture a money-loser in most cases. Two major mills established large-scale processing facilities in the burn, the most successful being the Carrier Lumber operation run by Bill Kordyban, an innovator with a reputation for profitably harvesting difficult-to-exploit stands.

Though he understands that his life may have been saved by a forest officer's ruse—in the form of a promise to protect his house during the evacuation—Lee Rose remains irked at having been duped into abandoning his home to the Swiss fire. Later he saw news footage shot by a BCTV film crew who had been at the ranch the day his family fled. Images shot shortly after the family's departure show the forest officer who had promised to protect the Rose house telling firefighters and camera crew to abandon the ranch as quickly as possible. The broadcast's final jerky images suggest that the camera operator had every reason to believe in the importance of a hasty retreat. Adding to that impression are images of husky loggers—Northwood's firefighting crews—running for their pickups with uncharacteristic speed.

The morning following the evacuation of Buck Flats, homeowners were told they would be allowed through the roadblock that had been set up to prevent all but emergency response vehicles from entering the fire area. The catch was that only male homeowners would be allowed back in. Dorothy Rose was ticked off about *that* slight, too. She was, by her own description, "healthy as a horse"; Lee was recovering from a heart attack. One of the Buck Flats women was also ticked off. She too was healthy, but her husband was recovering from leg surgery. She confronted the RCMP officer blocking the Buck Flats road. Getting no satisfaction, she rammed his police cruiser. He intended to arrest her, but changed his mind when a crowd of angry Buck Flats residents of both genders angrily demanded her release.

Lee returned to the ranch alone, his wife still forbidden to accompany him. Driving through Buck Flats he saw some neighbours' houses standing, others destroyed.

> Rose: The road up to our place was like driving through hell. Everything was absolutely black—the rocks, the ground, the trees. There were trees across our road but I didn't have a chain saw with me so I had to move them by hand. Fortunately they were all small. I couldn't do a lot of exertion because I was a heart patient. I drove up the last hill and everything was black, black, black. I got to the entrance to our property and the pasture was all green—Emerald City. I thought, Ah, our place is saved! The fire had veered around our property and left a ring of green trees practically around

the place. It wasn't until I turned into the driveway that I could see the house was gone and there was nothing left standing but the chimneys.

Picking through the ruins of his homestead, Lee found proof of the fire's intensity. Near the charred pen where a pair of pigs had burned to death, a partially full 450-kilogram propane tank had survived intact. Beneath it, several glass bottles were melted into wax-like distortions. Around the yard, Lee found fragments of a wooden windmill propeller whose tower had been located near the house. The propeller hadn't burned. It had disintegrated as it spun furiously in the fire-induced windstorm. Those pieces of propeller, as much as any other proof, convinced Lee that his initial conviction to stand and fight had been "a fantasy."

During the month the Rose family spent living in emergency accommodation in Houston, Lee got a phone call from a man who identified himself as a member of the Mennonite church. The church, the man said, was offering to rebuild the homes of any burned-out families who requested their help.

*The remains of the Rose homestead: "I drove up the last hill and everything was black, black, black," said Lee Rose.* BC Forest Service collection

"I'm sorry," Lee said, "but I'm not a member of the Mennonite church."

To which the called replied, "That doesn't matter. We just want to help."

Lee asked, "What do you expect in return? I'm not going to join the church."

He was told that Mennonite tradesmen were interested in providing help, not generating converts. In exchange for doing skilled labour and providing their own accommodation, breakfast and supper, church members asked that all building materials be provided and that they receive lunch on the job. Lee borrowed a flatbed truck and spent a month hauling supplies as Mennonites from Abbotsford, BC, and Vermilion, Alberta, cleared debris, built fences, planted the garden and resurrected a house on the site of the log home that had been razed.

Meanwhile, people in Houston organized a fundraising dance, with proceeds being divided among the fire's victims. Other kindnesses were forthcoming: a box of nails dropped off unbidden, additional free labour offered. One day a dark-visored motorcyclist wheeled into the ranch yard, handed Lee an envelope and rode away. The envelope contained $200.

Though their fire losses challenged the Rose family in several ways, the trial was not unrewarded. Sitting with his wife at their kitchen table—the table that "one smart man" had removed at Dorothy's request—Lee commented, "We had fallen into a nonchalant kind of life. The fire was a struggle that not only united Dorothy and I, it brought us closer to the whole town. It was like all of the neighbours that got burned out were one big family. We got very close. Some have died and some have moved away, but we're good friends with the ones that are left. We went through something together."

Eleven years after their lives were permanently altered by the Swiss fire, nine Buck Flats families received compensatory cheques from the Ministry of Forests. Six homes had been destroyed in the fire, along with farm machinery, barns and other structures. Only one house had been even partially insured.

A lone fire victim had been convinced that the compensation battle could be won. Sophia Ebermann was not inclined to give up without pursuing the cause to the bitter end, even when her neighbours had abandoned it. As Dorothy Rose recalled, "Everybody was saying, 'Sophia, it's no use.' And she said, 'I'm going to do it.' And she did."

Those who lost homes had maintained that, on the basis of wind and fire patterns, Forest Service backfires lit on the afternoon of May 30 may have been responsible for flames sweeping across the Rose ranch and into Buck Flats. In the end, financial reimbursement came as the result of an investigation by BC ombudsman Dulcie McCallum, who found that victims' complaints about the suppression effort had been mishandled by the Ministry of Forests "and [they] deserved compensation on that basis." Though the Ministry of Forests had eventually conducted an internal review of the Swiss suppression effort, McCallum found that the audit was too little, too late, but she also noted that her office found no "conclusive" evidence that the ministry's efforts to control the fire had directly or indirectly caused the Buck Flats or Rose losses.

To this day no Swiss fire victims know what their neighbours received in the financial settlement. Though the Forest Service distributed cheques totalling nearly a half-million dollars, the agreement required recipients to keep their payouts secret, ostensibly to minimize competitiveness or bad feelings.

When the Ebermann family—Frank, Sophia and six-year-old daughter Amai—had fled their Buck Flats home on the afternoon of May 30, they departed with a pickup-load of family possessions but left behind a dream in the form of the log home that Frank and Sophia had spent four years building. Though unfinished, the house was sufficiently complete that the family had moved in the previous winter. They had made everything in it themselves, including the wine stocked in their cellar.

Before being forced to leave, the Ebermanns had sat for a time on their lawn, watching the fire advance. Amai had packed a teddy bear and the family's photo albums. ("I was a very serious six-year-old," she recalls.) While she waited to leave she sat and drew a picture of their house. She drew sparks flying through the air. The picture, she told her mother, was of their house burning. The house isn't going to burn, her mother told her. The house is going to burn, Amai replied.

The Ebermanns were evacuated to the townhouse complex in Houston where they lived alongside the people who hours before had been their neighbours on Buck Flats Road. The families were largely sleepless that night as they contemplated the fate of their homesteads. The next morning fire officials informed the Ebermanns that their house had burned just after midnight. It was May 31, Sophia's birthday.

When the Ebermann family returned to the remains of their home they found that a firefighter had rescued Waldie, their miniature dachshund, which had been left behind during the evacuation. The low-slung dog had apparently been forced to jump over burning logs during the ordeal, as all the hair was singed off its stomach. Amai's search for a second dog, a lab named Blackie, ended under a burned-out cabin on the property where she found its remains alongside those of its five pups. That discovery, she guessed, was at the root of the fire nightmares she experienced for years after.

Frank Ebermann noted 19 years after the Swiss fire that his life "changed completely" the day he learned that his home and possessions had burned. "I certainly became less materialistic, less future-oriented. What was the point of gathering up material things when it could all be gone with one match?" Like Lee Rose, Ebermann hadn't believed that the fire would reach his house. And like Lee Rose, he had been promised that the place would be protected if it did. On the day after the fire the 38-year-old Ebermann had gathered with neighbours at the RCMP roadblock on Buck Flats Road. Among the people with whom he exchanged news were Heather and Tim Richet. Their house, they discovered, had been spared. A reporter with the newspaper *Houston Today* described the scene as Heather Richet attempted to divert Ebermann with conversation.

> The German who came to Canada eight years ago remained morose. His eyes were red from smoke and emotion. His hands were stuffed in the pockets of his jean cut-offs. On his T-shirt it read: "Life is a playground. Take your best swing."

Mennonite volunteers built Frank and Sophia Ebermann a new house on the Buck Flats land where their hand-built log home once stood. Amai—now Amai Barden—lives in Houston.

# CHAPTER 12

# Percy: Portrait of a Firefighter

In 1958 the BC Forest Service created a forest district centred on the southern Interior community of Ashcroft. John Noble, then ranger of the Barriere Forest District north of Kamloops, was selected to oversee the new administrative entity.

Noble had the rare quality of being admired by both staff and superiors, and his stature was enhanced by his respected family background. His father, Bill Noble, had made a 35-year career in forestry, based at Birch Island, near Clearwater, and had played an important role in the establishment of Wells Gray Provincial Park. Bill Noble retired in April 1944; two months later son John signed on for the first of his own 35 years of Forest Service employment.

Having taken charge at Ashcroft, one of Noble's first acts was to assess his roster of honorary fire wardens. Since the Ashcroft district had been carved from the existing Lillooet, Clinton and Merritt forest districts, and since Noble was well acquainted with Clinton ranger Charlie Robertson, it was there that he went for advice.

Noble: I went to see Charlie, to ask who the fire wardens were—what he thought of them, should I keep them on or not. Charlie said, "There's a young Indian lad down there by the name of Percy Minnabarriet. I always suspected that he might just have been setting fires. But I think you should get him as a fire warden because he has a lot of pride, and if he's a fire warden and his job is to see that there *aren't* fires, this will solve a lot of your problems."

("What would I want to set fires for?" Minnabarriet is reported to have growled when the fire-setting rumour surfaced. "I've already got a job.")

Noble drove to the Bonaparte reserve at Cache Creek—Minnabarriet had married into the Bonaparte band—and asked if he would consider becoming a fire warden. Minnabarriet was never a man to waste words. His response, as Noble remembered it, was "Fine."

*Percy Minnabarriet (lower left) with several cowboy friends. Minnabarriet's passion for the rodeo life would have conflicted with his firefighting career had he not established a close and mutually beneficial working relationship with Ashcroft forest ranger John Noble.* Photograph courtesy the Minnabarriet family

*Ashcroft forest ranger John Noble flanked by an assistant ranger (l) and ranger supervisor Don DeWitt (r). It was by "puppy-dogging" after Minnabarriet that Noble learned how to fight fire in the dry Interior.* Lloyd Siver photograph

The professional relationship founded on that simple agreement was one that would last for two decades, its success apparently based on the fact that it united two people whose mutual respect bridged what could have been an overwhelming cultural gulf. In their own worlds, Noble and Minnabarriet had proven themselves to be people of considerable power and talent. Together, the ranger soon realized, he and the man who had been presented to him as a possible arsonist could be a potent force to fight fire.

There's an expression among old-timers in Ashcroft that here even the rocks burn. This is a place where potential evaporation consistently exceeds precipitation, country well suited to the natural propagation of sagebrush and rabbitbush and rattlesnakes and fire.

John Noble had little dry-belt fire experience when he took over as Ashcroft's inaugural forest ranger, and he set out to rectify that shortcoming. Recognizing in Percy Minnabarriet someone who understood fire at a profound level, Noble began tagging after him on fires—"puppy-dogging," he called it. The ranger asked questions and the warden gave straightforward answers. For example, Minnabarriet lit backfires in conditions that would unnerve the most experienced of hands. Noble asked him why. "You're going to lose the fire anyway," he responded. "You might as well backfire and pick a good hot time of the day when it's going to go."

Noble was soon hiring Minnabarriet regularly as a fire foreman. From there he took the revolutionary step of assigning him to teach firefighting to new assistant rangers. First Nations communities had for decades provided some of the most talented and dependable firefighters in many parts of BC, but it was an almost exceptionless rule that the white guys in green uniforms did the talking and the Natives did the listening—and the physical labour. Now Noble had turned the tables. How well things went depends on who told the story. Noble remembered "one or two" of his underlings being the sort who, upon learning of their planned apprenticeship under Minnabarriet, replied, "No way am I working for an Indian." Minnabarriet's memory of the experiment was less upbeat.

*Percy Minnabarriet giving firefighting instructions to an assistant ranger at the Lytton airport, 1966.* Lloyd Siver photograph

Minnabarriet: Noble, he was a coyote, that guy, sending down these assistant rangers to me, telling me, "Take that guy with you. Show him how to fight fire." Me—an Indian. And as far as education, I've got piss-all. For me to take these guys out and tell 'em what to do, it didn't go over worth a shit with lots of them. I'd get lots of static.

If there was any compensation for the static Minnabarriet took from certain junior forest officers, it lay in Noble's refusal to indulge racism. Ashcroft was an ethnic melting pot and Noble mixed his fire crews without consideration for accent or skin colour. His humane and flexible management style notwithstanding, he expected compliance from his subordinates. The bottom line was that the ranger saw to it that Minnabarriet stayed on his payroll and intolerant staff members did not.

Dave Piggin was one of the assistant rangers Noble assigned to Minnabarriet's care. Now a forest health specialist in Kamloops, Piggin had begun his Forest Service career in Clearwater before being transferred to sun-baked Ashcroft.

> Piggin: Of all the firefighters that I've known, Percy is the person that I have the greatest respect for. He taught me how to fight fires. I'd heard his name for a few months before I met him because I started at the Ashcroft district in February, before fire season. And he was always spoken of with awe.

One business Minnabarriet excelled at outside of firefighting was rodeoing, but with summer being both rodeo and fire season, there was abundant potential for the two activities to conflict. Noble's respect for Minnabarriet was such that the ranger went out of his way to accommodate his fire foreman's passion.

> Noble: When I hired Percy as a fire foreman and we had a fire, I'd check to see when the rodeos were on. If it was just coming up to a weekend, I'd get someone else on so that Percy could take in his rodeo. He would come back on Monday and say, "I hear you've got a fire going. Do you need me?" And I would put him out there then.
>
> I remember one time Percy had a rodeo coming up and he was out on a fire. This was a Tuesday, and to get to the rodeo he would have to leave by Thursday. That night in the fire camp, when all the guys were sitting around the campfire, Percy said, "You guys know I've got a rodeo on Friday. This damn fire has to be cold-out by Thursday morning." It was. They'd do that for him.

I think Percy would have been there for me in spite of the rodeo if I'd asked him. I think he would have given up his rodeo if I'd said I really needed him. He might not have done that for everybody.

The mutual respect between ranger and fire foreman manifested again when Noble learned that one member of the Bonaparte band could beat Minnabarriet in certain rodeo events. It was a potential complication that the ranger adroitly accommodated.

> Noble: There was one Native in the band who would some-times beat Percy—he was better in some lines of rodeo work. Other than that, anybody in the Bonaparte band would work for Percy. The other fella, the one that beat Percy, I never *ever* put him on a fire crew with Percy. Because of their protocol, that wouldn't have been appropriate. I'd been told about this, so I never allowed it to happen.

Minnabarriet eventually inherited, and for a time moved his family onto, the 89 Mile Ranch near Spences Bridge that his great-grandmother had once owned. When fires broke out, Noble occasionally went there to pick up his foreman. Arriving one evening in his car, he met Marie, Percy's wife. "Percy's not here," she told Noble. Then she added, "Of course, he said that if *you* needed him you could find him along the river." Being "along the river" was code for the fact that Percy was dip-netting salmon in the traditional Native way. Until he gained legal Native status in 1987, he was governed by the same rules that applied to non-Natives. This didn't stop him from catching salmon; he just had to be careful about how he did it. With a houseful of children, salmon fishing was a matter of survival, not sport.

> Noble: I went and got a helicopter and flew down along the river. People just kind of disappeared when they saw us coming. They just melted into the rocks—they didn't know whether it was Fisheries or not. We landed and I hollered. Percy came out and agreed to come to the fire. He handed the car keys to one of his sons, got in the helicopter and away we went.

Dave Piggin recalled a time that a group of Forest Service employees contracted with Minnabarriet for a batch of 20 salmon. This was done on the hush-hush, the transaction being illegal anyway and the Forestry boys being upholders of the law besides. After the deal was done, the buyers noticed that one fish in the bunch was a scrawny little excuse for a salmon. They were briefly outraged, thinking they'd been shafted. Then they counted and found 21 fish: 20 prime Fraser River turkeys and one runt. Said Piggin, "He knew we were going to find that little one first and figure we'd been screwed. That was part of Percy's sense of humour."

*Dave Piggin worked with Minnabarriet for the best part of a decade. The renowned Native fire boss, he recalled, "was always spoken of with awe."* Keith Keller photograph

Minnabarriet's irreverence displayed itself once again when he and Piggin found themselves directing the suppression effort on a fire that had broken out on Nicoelten Mountain, north of Spences Bridge. True to his nature, John Noble had approved the hiring of a fire crew drawn from a nearby Hare Krishna commune. "They were very hard workers," Piggin recalled, "very easy to get along with. Tell them to build a fireguard from here to there, they built a fireguard from here to there." Minnabarriet too was impressed with the crew's diligence but couldn't resist referring to them collectively as "the Hairy Christians."

Until 1970 Minnabarriet's work with the Ashcroft Forest District was on an on-call basis. Noble resolved to change that. He had just assigned a summer suppression crew to Ashcroft, 10 seasonal firefighters ready to jump on new blazes the moment a fire report came in. He asked Minnabarriet to be his suppression crew foreman. "I can't do the paperwork on that," Minnabarriet replied bluntly. Those must have been difficult words, considering what a full-time government paycheque would mean to him. Noble had prepared his counter-argument. "You don't have to worry about that," he replied. "You'll have university kids on your crew. You can appoint one of them as your straw boss. *He* will be responsible for the paperwork. *You* fight the fires." Minnabarriet looked the ranger in the eye and told him once again, "Fine." Noble described how the experiment developed.

> Noble: The suppression crew that Percy ran—we used mostly university students—kept coming back to Ashcroft. And back and back and back. They liked working for him. There wasn't a man on his suppression crew that didn't know how to work a power saw, work a pump, drive a truck, use every tool properly. Every man on his crew knew how to do every job. It really paid off. One time we had 26 fires start in a four-hour period, and we were able to put out each of his men as a fire foreman. Percy had them trained that well. Loggers and everyone else accepted them as fire foremen; they never even questioned it. He had a real knack of getting the best out of everybody.
>
> My father taught me that you have to learn to delegate work, that it doesn't really matter how something is done, it's the results that pay off. Percy was very much that way with his suppression crew boys. He didn't tell them, he showed them and asked them. He was an exceptional man for that type of work.

Walking into fires, Minnabarriet had a powerful but unhurried pace. This was partly because in the late 1950s he'd had his left foot and leg crushed in a bulldozer track when his boot lace caught in the cleats. They'd managed to keep him in hospital at Kamloops for a year or so, but after he got out he repeatedly frustrated his doctor by cutting his hip-length cast to below the knee so he could get back to riding horses. Finally the doctor sealed him in steel rods, but they only lasted until Minnabarriet had Marie pick him up a new blade for his hacksaw. The other reason for his slow pace was that he engaged all his senses while walking to a fire. Once in a while, as his suppression boys hurried through the forest in what they hoped was the right direction, Minnabarriet would pick up a length of wood and explain to them that it was an Indian fire stick. He'd put one end to his ear and the other against a tree and concentrate hard. Then he'd walk to the fire. He'd actually been picking up clues—sights, sounds, smells—that everyone else was missing in their haste, though there is reportedly still at least one person working for the Forest Service who believes there just might have been some kind of Native voodoo in that stick thing after all.

Toward the end of the 1970s, Noble's superiors in Kamloops decided that under Minnabarriet's tutelage Ashcroft's suppression crew had become too good to remain intact. This decision was taken in the midst of a major Forest Service restructuring, a controversial organizational overhaul that eliminated rangers and ranger districts in favour of a more centralized, professionalized and bureaucratized system. In the opinion of many forest officers of Noble's generation, the era of a cohesive and effective Forest Service was about to close. Noble was grateful to be within a year of taking his retirement and, as he put it, "running for it."

At any rate, the Kamloops brass announced that they would be splitting up Ashcroft's summer crew and sending them in pairs to districts whose crews had evolved without the benefit of being formed by Percy Minnabarriet. Two of his crew members—one in his last year of medical school, the other about to become a lawyer—were directed by Kamloops to report to Merritt. They replied that they would go to Ashcroft and fight fires with Minnabarriet or not fight fires at all. Neither side would blink, which put an end to two firefighting careers one season before their time.

In developing his fire crews Minnabarriet preferred working with green city kids rather than their rural counterparts. The urban crowd, he said, "didn't know nothin' and they were willing to learn. They were willing to listen to you. But you get the guys from up around Prince George, kids from the Interior, it was always, 'My dad did this, my dad did that.' They'd cut firewood at home so they thought they knew something. Sure, maybe dad cuts two trees a year. And the kid cuts that tree up in blocks for firewood. So he figures he knows how to run a power saw. You had a hell of a time." Two rural teenagers who Minnabarriet did admire were Bruce and Ross Noble, John's sons. As a ranger, Noble was not allowed to hire family members; Minnabarriet would wait until Noble went away on summer holidays, then put the two boys on his crews. Both now fight fire for the BC Forest Service, Ross as an air attack officer, Bruce as superintendent of operations for the provincial air tanker program.

# — *Basque Roots* —

Louis Antoine Minnabarriet was the son of an aristocrat of French Basque ancestry. He left France in 1849 and began ranching in Oregon before setting out with a group of other Basque men on a gold-prospecting expedition to the Thompson River in 1858. Though he initially found little gold, Antoine and several partners did establish the renowned Basque Ranch, north of Spences Bridge—Cook's Ferry, as it was known then. He took a "country wife," Mary Lostair of the Cook's Ferry band, and the pair had a son, Louis Joseph Globe Minnabarriet, before Antoine replaced Mary with a white wife. Mary later married a man by the name of Peter Audap, owner of the 89 Mile Ranch and roadhouse on the Fraser River between Spences Bridge and Ashcroft. In the 1880s Antoine belatedly struck gold on the Thompson, sold the 800-hectare Basque Ranch—which he now owned outright—and returned to France with his white family. Louis, who had worked for his father as a blacksmith after being educated in New Westminster, moved to his mother's place at 89 Mile and went on to have a family of his own. Percy Minnabarriet was his grandson. Percy spent much of his life on ranches. His first job was on the Ashcroft Ranch, and he put in 15 years cowboying at the Upper Hat Creek Ranch. He even managed his great-grandfather's Basque Ranch briefly when Bethlehem Copper bought that property for the Bonaparte Native band in exchange for the right to dump mine tailings on reserve lands elsewhere. Ranching and rodeoing went hand in hand for Minnabarriet. At 16 he won $1.80 in his first steer-riding competition at Clinton, and during his quarter century on the professional circuit he won many other honours in competitions as far flung as Calgary and Cheyenne, Wyoming.

If Minnabarriet preferred the blank slate that urban recruits represented, it was because he was painfully aware of the responsibilities his role carried. His job required that he produce competent, decisive firefighters and that he bring them safely home at day's end.

> Minnabarriet: Not everyone can take the challenge of being the supervisor, the boss. Maybe they read the wrong book or some damn thing. You've got to depend on your crew. You've got to believe in your crew. You've got to be able to tell them, "This is where I want you to go. This is what I want you to do." You have to trust your guys and they have to trust you. One of the main things in firefighting is to know where your men are. You make sure you don't put them in a position where it's going to scare the shit out of 'em. When you're in charge of men you've always got to tell them, "You guys be careful. You've got to make sure you've got a way to get out. You've *got* to have a way out."

The crews that assembled in Ashcroft each summer consisted of keen young men, most of whom had passed the previous months honing their minds in lecture halls and libraries. Suddenly they had an assignment that didn't include books. They were being handed pumps and pickups, chain saws and Pulaskis and told they would soon be engaged in a form of potentially deadly combat with the most potent of elements. It is probable that heroic images formed in some of those minds—the urge to heroism being one of the motivating forces that impel young men into battle. Before adrenaline could combine with testosterone to produce disaster, Minnabarriet gathered his boys together and gave them a fatherly dose of reality.

> Minnabarriet: The thing I used to tell my suppression crews and tell all my firefighters—and it would probably be a harsh way of putting it—is: "I'm your superior, and what I tell you to do, you do. And you *listen* to me. You go out there and get hit on the head with a limb or the Cat pushes something on top of you—you get killed. For a few minutes everybody says, "That poor bugger." But tomorrow

morning it's a different story. They'll say, "If that dumb bug-
ger wasn't so dumb, he'd still be running around."

I said, "You listen. You break a shovel handle, you chuck
it in the bush and go get another one. That's exactly what
you are. You're just a tool."

Which is true.

If further convincing proved necessary, Minnabarriet had stories of fires gone
wrong. One involved a friend, a catskinner he worked with on a fire at Lac Le
Jeune in the 1950s—squashed in his seat when a tree came back on him.
Then there were the two guys at Lytton who got on the wrong end of a fire.
Minnabarriet was rodeoing at the Penticton Peach Festival when the disaster
occurred outside of Lytton in the late 1960s.

> Minnabarriet: At Lytton you sometimes get tremendous
> winds within a matter of minutes on account of the high
> mountains and the valleys. This one time it was hot, and the
> fire started going uphill. The Forestry guy sent two Indians
> up to the head end to cut the fire off—young guys, 18, 19.
> Some of them valleys are piled deep with fallen timber. And
> it's steep. They headed into this thing, then they had to run.
> They tried to run. But how are you going to run? Where are
> you going to go? They got cooked before they got out.

(It is a measure of John Noble's character that he in no way denies any part
he may have played in the deaths of those two young men. He cuts in
brusquely when it is put to him that a newly recruited assistant ranger was
in fact working that fatal fire: "*I* was the ranger." West of the Fraser River,
Noble said, he had a policy of keeping crews away from the top end of fires
until after sundown, at which time a downdraft invariably began blowing
off the surrounding mountains, allowing crews to move in safely. That's the
way the local old-timers had always fought fires, and that's what the survivors of the fatal blow-up testified at the inquest: "We were doing what
we've always done." In this case, a five-man crew waited until the downdraft
began before moving above the fire—just before an 80-kilometre-per-hour
wind sluiced up the Fraser Canyon and, in Noble's words, "the mountain

literally exploded." Among the three survivors, two got off the mountain immediately; the third came home the next day, having found himself two valleys over before he stopped running. Noble continued to use his sundown policy after that event, but he went to greater pains to get accurate spot weather forecasts before sending people ahead of fires.)

It was a point of great pride for Minnabarriet that he never lost a man on a fire. There was plenty of potential; he fought some fearsome fires. There was the 1970 Burtot, where he commanded nearly 300 men as flames closed in on the village of Lytton. Another one remarkable for its volatility cut loose in the Botanie Valley.

> Minnabarriet: In Botanie, way up on top, there's a place they call the Ruddick's Dam. The smoke went into a funnel, like a pipe. The smoke started spinning, then the fire went up the middle of it. Then the pipe bowed over to the other side of the valley—about five miles—and we had another huge fire over there. That was a big fire. Steep. It went furiously. There were big sparks—chunks of bark flying and landing in Lytton—and a bunch of people running around and putting them out.

Most firefighters had sense enough to know that they had been placed in the best possible hands when they were assigned to work under Minnabarriet. A few didn't. In one instance he encountered a crew that earned the distinction of being the worst bunch he ever ran.

> Minnabarriet: They were from Montreal, a bunch of Frenchmen that could hardly speak English. They wouldn't work for me. They gave me a bad time, sassing me back. So I called old Noble on the radio. I said, "I'm sending these guys back down to you tomorrow." I told them, "You can go home in the morning." And you know what the dirty bastards did? We were on a fire where we had to haul water in for 20 miles by pickup—water to cook with, to drink. We'd just got through bringing water up. A guy's gone down at three o'clock in the morning to bring it up over rough

roads, a Cat road made with one sweep of the Cat. And those guys took their dirty undershorts off before they left and stuffed them in the goddamn water cans. Our cook wasn't that old, he'd just got out of the army. And when he seen that, them shorts in that water tank, them Frenchmen didn't know how close they came to getting killed.

Those Frenchmen may have been the worst firefighters Minnabarriet ever ran into, but they weren't the only people to incur his wrath. He was supremely confident of his own skills, defending his own strategic choices even when they clashed with conventional firefighting wisdom. As he once commented with indignation evident in his voice, "I got into some terrible arguments with some [Forest Service personnel] who didn't know who I was." To the observation that he appeared to have had the complete confidence of staff who *did* know who he was, Minnabarriet paused, then humbly continued, "Other than an assistant ranger or a ranger, no one was allowed to call in a bomber on a fire. I could."

Noble, in fact, gave Minnabarriet the freedom to do pretty much whatever he felt was necessary to accomplish his job. When Minnabarriet wanted helicopters, he got them; he used helicopters quite a bit. "Old Noble," Minnabarriet said, "he was good. Whatever I needed, he would get it for me. 'Cause him and I were together lots. I believed in him and he believed in me." A fire in the headwaters of Izman Creek, north of Lytton, illustrated the confidence that existed between the ranger and his suppression foreman. Noble and one of his Kamloops superiors had stopped by around lunchtime to see how Minnabarriet was doing with the blaze.

Noble: About two o'clock in the afternoon Percy started moving all the Cats from one part of the fire down into another area that was quite quiet. And moving extra crews down in there too. My supervisor said, "What's he doing moving the crews down there? The hot spot's up on top."

I said, "Don't you worry about it. If Percy's moving the crews down there, he knows something." And boy, about four o'clock that area blew up. And he had the crews already there—no problem at all. And up on top, where it had been hot, it cooled down. I asked him after, "Why?"

He said, "John, I don't know. I just felt that there was going to be trouble, and if we didn't move our crews in there we were going to have to move our camp."

Even *he* didn't know why he did it, he just knew it had to be done. And he had guts enough to go ahead and do it. A lot of people *think* something should be done, but they don't want to be caught out in left field. That never bothered Percy.

In my mind I often thank Charlie Robertson for recommending Percy to me. Percy probably taught me about 90 percent of the firefighting that I knew in the dry belt. Protection staff used to say that I was really good at firefighting, but I was a good student. I watched what Percy did. Percy and I had a really good working relationship, a special working relationship. He was a fantastic firefighter, a fantastic man.

Percy Minnabarriet died in the spring of 2001. The Drylands Arena in Ashcroft was the only local building large enough to accommodate the funeral. Among the hundreds of people in attendance were John Noble and many of the 20 or so children—now adults—that Percy and Marie had nurtured for various lengths of time while raising their own six sons, three daughters and two adopted daughters.

Dave Piggin delivered the eulogy. He told the people gathered there how as teenagers Percy and his lifelong friend Poncho Wilson would chase wild horses in the Highland Valley, using them to practise their riding and roping skills. He told them how, when Percy began courting Marie, he would ride his horse cross-country to Cache Creek from the Upper Hat Creek Ranch, and how the couple would catch a bus to Ashcroft, watch a movie, eat supper at Bloom's Cafe, then catch the last bus back to Cache Creek so that Percy could saddle back up and ride home before daylight. He referred to the time Percy caught a "hippie" who'd "borrowed" his horse at the Williams Lake rodeo, and how Percy tracked the guy down, got a rope on him and managed to persuade him to not do that again. He told how, while the family lived for a time up at Venables Lake, between Cache Creek and Spences Bridge, Percy would light bonfires on the ice in wintertime so they could play hockey by firelight. He said that Percy had been "a legend in his own time" when it came to fighting fire. He mentioned quite a few other things, and then everyone drove in a procession down Highway 1 and buried Percy in the family cemetery at 89 Mile, in a grave overlooking Spatsum where he was born in a log cabin that doesn't exist any more. Between the places of his birth and burial the South Thompson River flows, powerful but unhurried.

*Whether it was hunting, fishing, ranching or firefighting, Percy Minnabarriet lived much of his life outdoors. He had an intimate knowledge of the rugged South Thompson River country.* Photograph courtesy the Minnabarriet family

# CHAPTER 13

# The Summer of Their Discontent

In 1985 when the Ministry of Forests recovered from the immediate shock of that year's brutal fire season, it convened boards of review to assess the province's nine worst fires and the efforts mounted to put them out. The executive summary those boards produced opened with the statement, "1985 was regarded as the worst fire season in the 74-year history of the British Columbia Ministry of Forests." It was a powerful contention, though an arguable one, there being several possible ways of determining the worst fire season: by the number of fires, the area burned, the amount spent on suppression. However, it was likely the worst season that most people working for the Forest Service at that time had lived through. And it may have been one of the uglier seasons the Forest Service faced internally, a summer of organizational discontent. In addition, considering the changes instituted after the subsequent period of internal and external criticism, 1985 may also be the best candidate for the watershed year in which BC's forest-firefighting machinery entered a new, modern era.

## The Summer of Their Discontent (Invermere Fires, 1985)

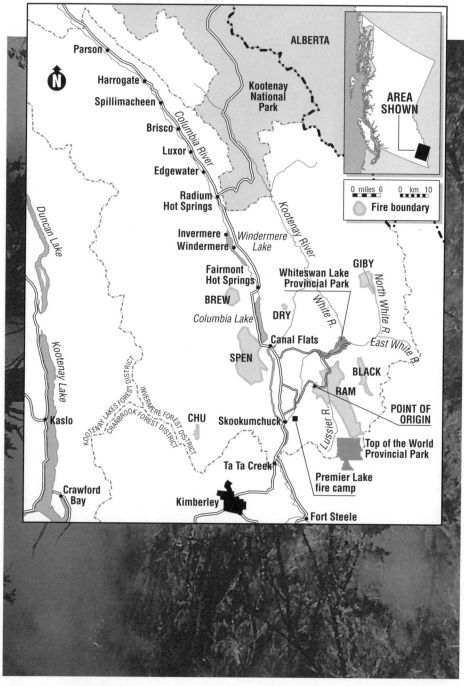

Spring 1985 had arrived dry and cool in the Columbia Valley. The dry persisted as spring progressed; the cool did not. By the eve of the July long weekend, with summer just a week old, the runoff had already peaked, soil moisture was well below normal and the fire danger rating had climbed to high. Still, the Invermere Forest District went into that holiday weekend having experienced a mere 20 fires and having controlled them at an average cost of under $900. The district had a six-man initial attack team available and was sending up spotter planes on an as-needed basis. Things seemed to be under control.

Then that Friday night a dry lightning storm surged through southeastern BC, sparking a major fire bust that would, over the next two-and-a-half fiery weeks, try the BC Forest Service in a way it has rarely been tested. Nearly 1,700 firefighters eventually signed onto the Forest Service payroll, their efforts overseen by 160 hastily assembled government personnel and backed by 40 helicopters and dozens of pieces of heavy equipment. Most of the firefighters were untrained volunteers, soldiers of fortune who flooded into Invermere from across Canada, drawn by the prospect of a paycheque and the excitement generated by an uncontained natural disaster.

Two fires contributed disproportionately to the more than $10 million that taxpayers eventually spent on the Invermere battles. The Spen, named for its point of origin near Spencer Creek, forced the controversial evacuation of Canal Flats and drew the lion's share of media attention. It was a made-for-TV fire, with dramatic activity on the outskirts of town, helicopters dropping through smoke to bucket water on endangered houses, and angry residents vociferously protesting their forced departure. In addition, there was an apocalyptic prediction that several major blazes might merge and generate a lethal oxygen-consuming firestorm. In the end the town was saved and undercover police posing as firefighters arrested and charged with arson a Canal Flats resident who had been hired to fight the fire that he had set.

The second major blaze began with a vague report of smoke issuing from a nearly inaccessible cliff at Ram Creek Hot Springs, high in a Rocky Mountain pass 65 kilometres south of Invermere. Greg Pommier, logging planning supervisor for Crestbrook Forest Industries, hiked into that high country early the next morning to flag a route to the fire. Having evaluated the situation, he envisioned how the suppression effort should proceed: run guards up either side of the fire to corner it in the high, rocky country, then move in with air-supported ground crews to finish the thing off. Instead, he had to watch as "it grew into a monster," the result of a crippling combination of human and natural factors. First there was the weather—this rash of lightning strikes had arrived just before the area's fire hazard index hit extreme. Being a long weekend, few people—either Forest Service personnel or loggers—were available for firefighting. Those who were on hand found themselves severely short of resources because the Ram began just as the district's first trouble fire, the Brew, was sucking up every firefighter and piece of equipment that could be rounded up, and only two days before the Spen ignited. So the Ram got what remained after other priorities were attended to—that is, practically nothing.

The man in charge of fire control for the Invermere district in 1985 was resource officer Dunc Hamilton. Given his position, he had to bear some responsibility for the Invermere bust developing the way it did, but colleagues suggest that he was assigned more than his share of the blame. "I think that Dunc got a dirty deal," is the enduring opinion of Ed Cunliffe, Invermere's head of log scaling in 1985 and an experienced fire hand. "He took most of the flack. They needed a goat—someone to pin the thing on—and I think they used Dunc." When Hamilton was interviewed about the firefighting aspects of his Forest Service career shortly before his death in 1998, he did not discuss Invermere in 1985.

With no one else to hand the Ram over to, Hamilton served as its first fire boss. He began by requesting a Rapattack crew—firefighters who rappel from helicopters into difficult country—to begin constructing heliports at strategic locations near the kilometre-long fire. He was told by regional headquarters that the chance of getting such help was "poor." It was. He never got it. He requested air tanker drops to slow the Ram's spread and was advised that air support would not be forthcoming "due to other commitments and severity of terrain." Lacking any other option, he had a crew hike in to determine the feasibility of foot access to the fire's hard-to-reach top end. He had to wait for the crew to return to hear their findings because—as would be the case for days to come—Ram firefighters had not been issued radios.

Over the next dozen days the fire slopped out of Ram Pass and began burning down toward the heavily timbered Lussier River Valley. Swamped by other responsibilities, Hamilton passed the fire over to Ed Cunliffe, who was joined on July 10 by a hastily assembled overhead team headed by Bill

Jackson, the person who had led the fight against the 1971 Tee fire. Jackson, Hamilton's counterpart in the Prince Rupert Forest District, was in Invermere on July 10 being briefed on the Ram situation when gusting winds and scorching temperatures blew the fire out of control on all flanks. The next day, when his team officially took over, the Ram had increased to 6,000 hectares, and over the following 24 hours it more than doubled again. To fight their monster the team had almost non-existent communications, two light helicopters, 100 firefighters and inadequate camp facilities.

> Jackson: There was nothing left in the province. We couldn't get radios, couldn't get helicopters, couldn't get camp stuff. I begged and pleaded and demanded for what I needed. We needed helicopters, we needed a big crew. This was in a major timber area—pretty valuable country—and industry was pretty upset about it.

*Jack Carradice at the Ram fire, 1985: "He's forceful, he's gung-ho, and anything that gets in his way gets cut down and pushed aside."*
Photograph courtesy Jack Carradice

A week later Jackson's firefighting force had increased to about 400. It was being fed from two 50-person cook trailers. They still had far less equipment than they needed and so few radios that firefighters were refusing to work on the Ram's more volatile sectors.

The Invermere Forest District had remained in charge of the surrounding fires until July 7. However, according to one Forest Service employee, "They didn't understand the logistics of handling the problem. They didn't understand that they were in *big* trouble. It took some time before that really sank in." When the scope of the crisis did sink in, the district turned control of its biggest fires over to the province, which in turn parachuted in a management team to kick-start the recovery effort. Among that team was Jack Carradice, the person who would develop the highest profile at Invermere—his face appearing on the nightly news looking by turns grim, determined, haggard, triumphant. He was manager of the Chilliwack Forest District when Protection Branch director Hank Doerksen phoned and gave him the standing instruction for such emergencies—see what you can do with it.

Few people who know or worked with Carradice are equivocal in their opinions of him. Some called him John Wayne; others called him other things. "Jack is flamboyant, that's his nature," commented a former colleague. "Things happen when Jack's around. He's very blunt, doesn't mince his words. He's very forceful, he's gung-ho, and anything that gets in his way gets cut down and pushed aside." As a forest ranger in Squamish in 1960 Carradice had laid formal charges against nine loggers who he felt weren't fighting a fire with due diligence. The resulting trial sparked an uproar among local members of the International Woodworkers of America and an accusation of incompetence and negligence against Carradice. "Strong loggers and a bull-headed ranger," he later said of that squabble. "It got to the point where they wouldn't bend and I wouldn't bend. It got to be a bit of a media circus."

A stint in a media circus was good preparation for Carradice in July 1985 because that's what he stepped into the day he arrived. Among many things the Forest Service learned that summer was that people are fascinated by big fires and want to know more about them. The media, of course, was already aware of this, and that is why there were dozens of reporters and photographers milling about when Carradice touched down at Invermere. Many were in a sour mood, having been pretty much shut out in their attempts to get information, interviews or a first-hand look at what the flap was all about. Not that there hadn't been patchwork attempts to provide the media with what they wanted, but Forest Service staff thrown into the breach were simply overwhelmed. A retired Ministry of Forests public relations director was eventually sent in to help placate the journalistic throng; overwhelmed, he resigned two days later. A hired "information services consultant" lasted one day before being relieved of her post. Finally Carradice had reporters create a pool and pick one representative who would feed material back to the other media types stranded at Invermere. The winner was CBC national television reporter Kathryn Wright, and from then on she and her cameraman rode in the back seat of Carradice's helicopter as he toured the war zone. Carradice let Wright in on everything—overhead team briefings, strategy sessions, the works. The only thing he asked was that she look the other way when it was necessary for him to deal with "confidential personal situations," which generally meant people coming unglued from stress or exhaustion or both. He also appealed for, and got, media silence when RCMP went undercover on Spen fire lines to catch the arsonist who'd lit that fire and who continued to light new ones as it burned. Overall, Carradice said, "I thought I got excellent treatment from the media. They were very responsible." There were some noises made in certain Forest Service circles about Carradice fancying himself a TV star. On the other hand, he didn't seem to be doing too badly, considering the circumstances.

If Carradice's supporters and detractors agree on one thing, it's that they wouldn't have wanted to be in his shoes that summer. One controversial action that he and fellow crisis managers took was the conversion of Invermere's high school to a centre of fire operations. Some field people resented the bureaucratic labyrinth that developed there. "The bureaucracy was just about big enough that it would have functioned on its own—without a fire," quipped fire boss Keith Blom. He was spending upwards of a hundred thousand dollars a day on his fire but wasted hours arguing for approval to buy a freezer to keep food from spoiling in camp. This wasn't necessarily Carradice's fault, but it was one of those things that contributed to the bad taste about what went on at the high school. "We had a name for the place but we never got around to putting the sign up," Blom said. "We referred to it as Malfunction Junction."

In his final report on the Ram fire Bill Jackson wrote in reference to the situation at David Thompson High School: "It became apparent very quickly that many people were in decision-making roles that did not fully comprehend the concept of firefighting and its needs. Some of the decisions which they made seriously and adversely affected firefighting activities." That shot was not directed at Carradice, Jackson later said, since he admired much in the Invermere boss's approach. "Jack knew what he had built, but it got out of hand. It wasn't all his doing."

Ed Cunliffe, who helped fight the Ram fire after Dunc Hamilton's departure, never lost what he described as "bad feelings about that whole mess," which he saw as being centred at the school.

> Cunliffe: That fire centre became a monster that got out of hand, with all that high-priced help. One guy would come in, then he'd phone his friend and bring *him* in. And it just kept going and going and going. It was ridiculous, in my opinion. If only all the resources and energy that was put into building that empire was put in the field. They had about six helicopters at the fire centre's disposal at one point. Those guys were flying all over the country, checking on everybody else and saying, "You should do this and you should do that." And then they'd fly back to the high school.

Cunliffe's ire stemmed in part from the time that Carradice and Tom Blom (Keith's brother), the province's superintendent of fire control, helicoptered into the Lussier River sector and ordered Cunliffe and CFI's Greg Pommier to light up a big stretch of the valley in an attempt to burn off fuels ahead of the approaching fire. The Ram was just beginning to break out of the pass where it had started, Carradice said, and had developed distinct southern and northern fronts. He and Blom had taken a "very aggressive stance" and had conducted what he considered to be a successful burn-off on the southern sector. He had suspected from the start that Cunliffe and Pommier didn't like the idea of a northern burn-off. They didn't. Pommier was one of the people who referred to Carradice as John Wayne, and though he meant it in an admiring way, he didn't want to do the burn-off any more than Cunliffe did. Still, an order was an order, so the two men fired up their drip torches and lit a line

*A DC-6 air tanker lays retardant in Quinn Creek Pass in preparation for a burn-off designed to stop the Ram fire escaping down Quinn Creek.*
Photograph courtesy Jack Carradice

along the main Lussier River road. The resulting fire didn't go so well that evening but it certainly took off in the next morning's heat, boiling up the valley toward Whiteswan Lake Provincial Park and off into a drainage known as Coyote Creek. Carradice speculated that the thing went wrong because Cunliffe and Pommier "were just a little conservative in the way they approached it and the fire beat them." That's not the way Pommier saw it.

> Pommier: They made a bad mistake there, bringing the fire down into the main valley bottom, into the old cut blocks and the reserve timber. The day we lit that the fire was about 1,000 hectares. I think the next day it was 6,000 and the day after that about 12,000. I remember being in a chopper one of those evenings with a Forest Service guy by the name of Leo Duffels, the line boss. And man, we just watched that fire going past those yellow kilometre signs on the side of the road.

# — *More Beer! More Soap!* —

From the Ministry of Forests report on the Invermere fires of 1985, detailing an event of Sunday, July 9:

> During the evening, firefighters from the Brew and Ram fires who had been housed at the David Thompson High School in Invermere rebelled when they returned and found they were to be camped there another night. Their major grievances were: lack of shower facilities, lack of sleep due to noise in the gymnasium and a directive that they were not allowed to go to town for a beer. The directive was revoked by Clint Nelson, who ordered that towels and soap be bought immediately for the men. Local merchants were contacted that Sunday evening and the shower supplies purchased and delivered. The men cleaned up and settled down.

From the July 17, 1985, edition of the Lake Windermere *Echo*:

> Joan Gross of Joan's Bath Boutique had a pleasant surprise last week when the Ministry of Forests purchased her entire supply of 120 bath towels for $1,500.

It is while reflecting on such situations that firefighters tend to bring up the old adage about hindsight being 20-20. They generally add that they did the best they could with the information they had at the time. That appears to have been the case with what turned into the failed attempt to head the Ram off at the pass. Exactly who had the better intuition about the Lussier River burn-off is difficult to decide at this distance. What is certain is that the fire had finally become entirely worthy of Pommier's "monster" designation.

It was at this point that Bill Jackson entered the scene. He had been eating smoke for about 30 years when he was pointed at the exploding Ram and told to do what he could with it. What he did—with the help of several hundred people—was bring it to its knees, a laudable feat given the terrain, the burning conditions and the initial shortages of just about everything. However, rather than coming away elated at his accomplishment, Jackson walked away "totally disillusioned," feeling that his reward for taking on this monumental challenge was persecution. He never fought another fire.

I first learned about the Ram from members of the overhead team who took it over from Jackson in its final stages. Having been summoned to help out with the Invermere emergency, two O-teams, one headed by Keith Blom, the other by Roy Benson, had been forced to decide which would tackle a fire in the isolated valley of Blackfoot Squaw Creek—the Black—and who would manage the higher visibility Brew fire near Lake Windermere. Both jobs brought their challenges. The Black was already at 500 hectares the afternoon of its discovery and by that evening was six kilometres long and roaring through its drainage "ridge to ridge." The Brew had blown out of control an hour after 10 firefighters began fighting it, incinerating four pickup trucks, a bulldozer and a portable sawmill.

Each O-team brought special expertise. Roy Benson, though born far from forests in the Alberta badland town of Drumheller, had, since joining the BC Forest Service in 1967, developed a reputation as a cool, astute fire tactician and overhead team leader. Keith Blom was a seasoned firefighter who had begun his career as an assistant ranger nearly two decades earlier in McBride—steep wet country where a typical fire involved a lightning strike on a hollow cedar snag. "Trench warfare is what it amounted to," he said. "It was a hell of a lot of work to fall cedar trees, buck 'em up and wash 'em out. And most of the time it was raining while you were doing it." Blom's air boss, Wayne Langlois, had grown up aspiring to fight fire as he tagged after his father, an honorary fire warden in Creston. He had lied about his age to be hired on his first blaze at 15, an event in which another 15-year-old died by being crushed under a bulldozer. Since then, he had gained 27 years of experience, and by the time he arrived at Invermere was considered one of the province's top firefighters. It was Benson and Langlois who tossed the coin to decide which team would take the Black and which would take the Brew. Langlois won the toss and, knowing the

Brew would be high profile and draw a lot of media and public attention, he chose the Black.

After a couple of intense weeks of work, the Blom O-team was pulled from the Black for a few days off. Then, having enjoyed their R&R, they were sent in to relieve Bill Jackson's team on the Ram. Along with Blom and Langlois were Jeff Berry, then a relative newcomer to the world of aerial firefighting and currently the director of BC's air tanker program, and administration boss Dave Procter, the person responsible for payroll on the fire as well as for every other accounting headache that comes with housing and feeding several hundred firefighters. No one on the incoming Blom O-team was overly impressed with what they found at Jackson's Premier Lake fire camp.

> Langlois: That first night, suppertime—we got there just before supper—there was people drinking already, drunks staggering around. A father and son got up on the table and had a fight. Right on the table. Both drunk. The father beat the hell out of this kid. That was our introduction to the camp. Talking to the other people on the fire, I find out there's undercover narcotics agents in the camp—a drug problem on top of the booze problem.

> Blom: They had everything in that camp—booze, drugs, hookers. There was people sleeping in until nine, ten in the morning. Didn't matter—they'd just fly them to the fire whenever they got up. There was no structure, no organization. And no accountability. It was just chaos. You've got to get it under control the first day, get it structured, get it running and keep it there. Whether you make friends or influence people doesn't matter.

> Berry: We had three undercover narcotics agents making regular drug busts.

> Langlois: The hookers were parked just outside of the area. They had a motorhome, I believe it was.

*This photograph, taken near noon on July 14, shows the main body of the Ram fire burning up Coyote Creek. The smoke on the right emanates from the newly lit burn-off at Quinn Creek Pass.* Photograph courtesy Jack Carradice

Procter: The hookers even took Mastercard and Chargex. They were on private land and the police couldn't do anything about it. They couldn't or they wouldn't.

Langlois: There was a bootlegger. He was a resident about a half kilometre from the fire camp. He had a couple of Doberman dogs, and even the police wouldn't go near the place. So there was a bootlegger there, prostitutes at the gate, booze in the camp. One of the questions they generally ask on a job interview is, "What would you do *if...*" This fire had every element.

I remember the first night or the second night. They had TVs or video machines in there, and somebody brought in one of these ninja movies. All night long, guys were out there kicking hell out of each other, all drunk. They cut up tents with knives, yelling and fighting.

Berry: The next morning we get up and one of the crews had cut a skylight and a back door in their tent. They'd been up drinking and partying all night, right beside the fire boss's tent—Blom's tent. He got up and said, "Berry, fire the whole lot of them!" So one of my first jobs was to fire the whole heliport crew. They'd been doing a good job up till then, but they certainly stepped over the line.

I likened it to Dawson City back in the gold rush. There were a ton of people there. It was chaotic. One day Wayne sent me up into the alpine to set up a burning station—we pre-mixed Alumagel in 45-gallon drums ready to be hooked on a helicopter, so the choppers could just drop down from the alpine to wherever it was they had to do a burn-off. I landed up there in a helicopter and people just scattered out of the alpine trees where they'd set up little camps for the day. They weren't even going down to the fire. They were flying people up to the alpine in a Sikorsky S-61 and these people would land and just bail out into the trees. There was a week's worth of lunch bags in there, hose boxes laid out for beds. I remember one day Keith [Blom] firing everybody that was clean that got off the helicopter at the end of the day. There were people coming in, landing at camp, rubbing the sleep out of their eyes, spotlessly clean. You can't fight a fire for 10 hours and come home clean. That was their last day.

Langlois: There was a helicopter pilot out of Alberta who told me, "You've just ruined the best time I ever had in my life." I said, "That's too bad. You get your helicopter and get the hell out of here." He got all his stuff packed and he hooked his bucket onto his helicopter. This was in a big field where they had the camp. I was near the middle of it, and he took off with the bucket slung underneath him. He came chasing me right across that field, trying to hit me with that bucket. Just ruined the best time he'd ever had.

We had to control that situation before we did much with the fire. Eventually we moved the camp. It had been right down by the highway, and the fire was 40 kilometres away. In the end we just took the camp and moved it right over to the fire.

I think they had 27 helicopters and 500 people at that camp, so we couldn't just jump in and take over right away. You've got to get familiar with things, like where *is* the fire? We started dealing with it right away, but we wanted to figure out what was going on and why. Why it was a wet camp instead of a dry camp. Why there was hookers and narcs. It didn't take any genius or any amount of time to see that things were out of control. We hurt a lot—*a lot*—of feelings, let a lot of people go. They just couldn't work under us. It was no longer a wet camp—now there was no booze at all. And we made representation to the RCMP to get those hookers and those bootleggers out of here.

From the time we took the Ram fire over it took us about two or three days to really straighten things out. There's a two- to three-day turnover period on a fire—so the other team was still there. The fire boss that was on there didn't want to give up that fire. He probably wanted to straighten things out before somebody else got in there. He was not a happy person when we took it over—amongst about 300 other people.

Jackson was certainly not a happy person either. Prior to the turnover his team had encountered some activities that they had "nipped in the bud," but he felt that there was nothing significantly wrong at the camp. He'd seen hookers set up shop near road-accessible fires before, and had found that moving them along was one thing but that getting rid of them altogether was quite another. Besides, he had other things to worry about.

Jackson: It's a fact of life that when you have a camp along a highway you are going to have problems. Unless you search the crew's personal effects, drugs and booze will always show up. A lot of people like to have a drink after a stressful day, and if we judge the people to be responsible we let it happen. If it gets out of hand you deal with it. You have to deal with each and every situation, taking into consideration the effect it will have on the morale and overall performance of your crew. Up to the time my team left the Ram fire, drinking was not a problem—it was happening but it was well under control. We let the odd unruly person go, if for no other reason than to make an example for the others. It would have been impossible to maintain a dry camp in this situation because we had all the equipment operators driving back and forth every day.

As with the Tee fire of 14 years earlier, Jackson had sensed that Ram crews were getting tired and ready for a blow-out, and by the time the Blom O-team arrived, they'd backlogged enough paycheques to afford one. Jackson surmised that the party atmosphere that erupted during the turnover period resulted from a combination of battle fatigue and his crew's dissatisfaction with the new team. The Blom team, he said, "was on a different mission. I don't know what they were up to. They moved in with preconceived notions and in a housecleaning mood. When we left about three-quarters of the crew left with us. They spent about a half a day with Blom and they left. It was the screaming management style. When someone loses their voice from hollering at their crew, there's something wrong."

So there is an element of interpretation involved here. Either the Ram camp was poorly managed and a lot of unruly firefighters had to go because they *couldn't* meet the incoming team's standards, or the Ram camp had run within tolerable limits and a lot of weary and cash-rich firefighters left because they *wouldn't* meet the incoming team's standards.

But if Jackson was unhappy when he turned the fire over to Keith Blom and company, he was livid that fall when he read the Forest Service report analysing actions taken on the Ram fire. The document he opened back at his Prince Rupert office was something that, he said, "blew my mind."

Jackson: A fire review is fine—you justify the costs and what went on. But these fire audits became a sort of head-hunting thing, a witch hunt. They would get the audit done by these foresters who were out on the golf course all summer long, people who never saw any smoke or flame. I sat in my office and read that report and I just dropped it in the garbage can. I couldn't believe what I was reading.

Jackson would later pick the report out of his round file and pursue the matter with other disgruntled Forest Service employees, among them Jack Carradice, who "didn't have a hell of a lot of respect for [the review]," because—and there was a distinct hint of John Wayne coming through as Carradice said it—"I thought there was a hell of a lot of back-bitin' bullshit in there."

Jackson: We controlled that fire under the worst conditions, brought it down to a mop-up stage when we left. We spent lots of money—yes. When we got started on that fire, I told Jack Carradice, "This is going to cost you money. If you don't spend the money on it, she's going to go out of the valley and down into the Cranbrook country."

One of the things the auditors said was that the camp at Premier Lake was in the wrong location. They said that more than once. Logistically, for accessing the fire, yes it was. Especially once the fire was under control. At that point we contemplated moving it, but the logistics of moving it were pretty prohibitive, and you've got to weigh that.

We put a camp in where we thought it was really safe. It was a logistic problem flying crews to work in the morning, but when you've got a 500-man camp you've got to think of escape routes and all those sorts of things. We had *very* volatile burning conditions, so we put the camp where there was a creek and a bit of a field for protection. As it turned out it was an excellent place. They even brought the army in from back east to do a camp inspection—to see how safe the camp was—and they were quite happy with it.

A few days after meeting with Jackson in Prince Rupert I met Dawson Wallin at his suburban Prince George home. Wallin, who had been recommended to me by a number of people, including Keith Blom and Wayne Langlois, had retired from the Forest Service in 1988 as protection officer for the Prince George forest region. Today he operates his own guide-outfitting business in the East Kootenay. People who worked with him described him to me as a talented firefighter and a savvy, fair-minded, straight shooter.

Wallin had major fires of his own to deal with in July 1985, but Prince George hadn't developed the profile the Kootenay bust had, so he was told to report there on the double. Then Carradice took all the Invermere fires and divided them down the middle, making Wallin responsible for half and another career Forest Service man responsible for the rest. Wallin's share included the Ram and the Black. Each fire boss was to report to Wallin, who would in turn report to Carradice. Wallin, reasonably enough, asked Carradice what he was supposed to do.

"We don't give a damn what you do," is how Wallin recalled Carradice's response. "Get the damn fires under control. Get it done."

Though Wallin oversaw some tough fires at Invermere, the Ram, for various reasons, stands out in his memory for the magnitude of the challenge it represented. Searching for the right word to describe that challenge, he settled on "astronomical." The fire covered 9,000 hectares when Wallin first arrived on the scene, and it was "going hell-west and crooked in every direction you could imagine." A camp was only beginning to be set up. After many representations by Jackson and Wallin, people and equipment began to pour in. All of this to be managed by Jackson and the other members of his pick-up overhead team, whom Wallin collectively referred to as "a few bodies just kind of thrown together."

> Wallin: They had their orders: "We don't care how you put the fire out. Just do it." And that fire was put out by the guys on the ground. Then, when the smoke all cleared, I think they decided to hang the wrong guys.
>
> When they did the audit, there was a lot of what I guess you'd call innuendo. For example, they said there were indiscriminate backfires that were lit and failed. And I guess there were some of those. I don't think any of those were lit

*The Ram fire and the Quinn Creek burn-off meet.*
Photograph courtesy Jack Carradice

when Bill Jackson was the fire boss. He did light up the whole end of one valley one afternoon. Log decks, Crestbrook's [Crestbrook Forest Industries] development work, wood piled all over—he burnt the whole goddamn works. It was the only thing he could do. And that stopped the fire from going into Quinn Creek, where it could have made a run down to Fort Steele, or gone around the corner and outflanked the Black fire. What Bill did worked, and it worked damn well.

The Ram fire was like having five major fires. Sure they spent $5 million fighting it. But the question I would ask is, "What value would you place on Top of the World Provincial Park?" That park could have burned easily. And Parks did *not* want Cats going into Top of the World Park—

# — *Paying the Bill* —

Glen Bertram was operations manager for the Chilliwack Forest District when he was called in to help manage the Invermere outbreak, but he was one of four people badly burned when his plane crashed as it was leaving Invermere. Before that accident he spent much of July organizing supplies and overseeing their transport to and from the myriad fires burning around the East Kootenay.

> Bertram: At first things were terrible, a horror show. Then Jack Carradice took control and things started to smooth out a little bit. Jack was a good organizer—he could organize things out of straight chaos.
>
> There was just so much going on, so many different fires. You'd ship stuff out but it would go to the wrong fire. Chaos. Finally we got a warehouse set up properly and somebody running it. They made all these fire bosses send orders to me, then I'd tell the warehouse what we needed.

no way. Bill had a big Cat guard built right tight against Top of the World Park, then he parked the Cats right at the boundary. And it worked. So how much is that park worth? Ten million dollars? Fifty million? If they did nothing else, I think that saving Top of the World Park was money well spent.

The Black was a dirty fire in tough terrain, but on a scale of one to 10, the Black was a two and the Ram was a 10. You can't compare what went on at those two fires. There's no similarities. When you get into a fire the size of the Ram, with its terrain and location and the number of men you've got, there's no way that a fire boss and an overhead team

I remember ordering foam retardant from some outfit in Montreal. I phoned them and said, "I need 600 pails of this stuff." He said, "When do you need it?"

"Tomorrow."

He said, "Tomorrow? This is going to take five days to get to you."

I said, "We've got fires burning everywhere. What's the best you can do?"

He said, "We can air freight it to Calgary and truck it to Invermere."

I said, "When will it get here?"

He said, "Six o'clock tomorrow morning."

"Good enough."

They chartered a 707 and flew it from Montreal to Calgary. I got to work at six o'clock that morning. At six-thirty a semi-trailer pulled up loaded with this stuff. I never did see the bill—I would like to have seen that. The chemical company chartered the aircraft, but you-know-who was paying the bill.

can manage the whole thing. In the initial stages there's going to be a whole lot of things where people are going to say, "This wasn't done right," and "that wasn't done right." There was easy highway access to that camp, so people could come and go. And when you've got about 400 guys together—and quite a hodge-podge of people—it's possible for some things to get out of hand at first. I think that all in all the stuff that went on there wasn't such a big deal. When things start to die down—that's when Bill left and Keith [Blom] and Wayne [Langlois] came in—then you start to get a handle on things. You start to get some semblance of law and order, of routine. Then you can start looking at safety things—guys riding on Cats, drugs being pushed in camp. You can start addressing those kinds of issues. In the initial stage—when that thing went from 9,000 hectares to 17,000 hectares—Bill Jackson had both pockets plumb full and he didn't need another jelly bean to stick in there. He was up to his ass in alligators.

It's like a war. When things start to simmer down, you get a handle on things. You start addressing the little things. But it's those little things that seem to bother people when they come in afterwards to do an audit. They don't seem to look at all the big good things that were done. They look at little things that didn't get done—a lot of nit-picking. All of those guys—Langlois, Blom, Jackson, Roy Benson, [Spen fire boss] Mike Cleaver—did an *outstanding* job that summer. That's what needed to be recognized.

It's easy to come along in November and say, "Why the hell was the camp way out at Premier Lake?" But if those auditors had been there in July when it was 30-something degrees and that sucker was sending a [smoke] column up to 30,000 feet every afternoon, maybe they'd have said that the camp should have been moved *further* away.

# CHAPTER 14

# Do You Believe in Prayer?

Red Deer Creek flows east out of the Hart Ranges in that transitional zone of northeastern BC where the Rockies, marching north to south, show they are no longer satisfied with being foothills and have every intention of becoming serious mountains. The creek originates on the spine of the Continental Divide, contributing its flow to that of the Wapiti River just short of the Alberta border, at which point the waters flow north in ever-broadening streams until they empty into the Arctic Ocean.

# Do You Believe in Prayer? (Red Deer Creek, 1987)

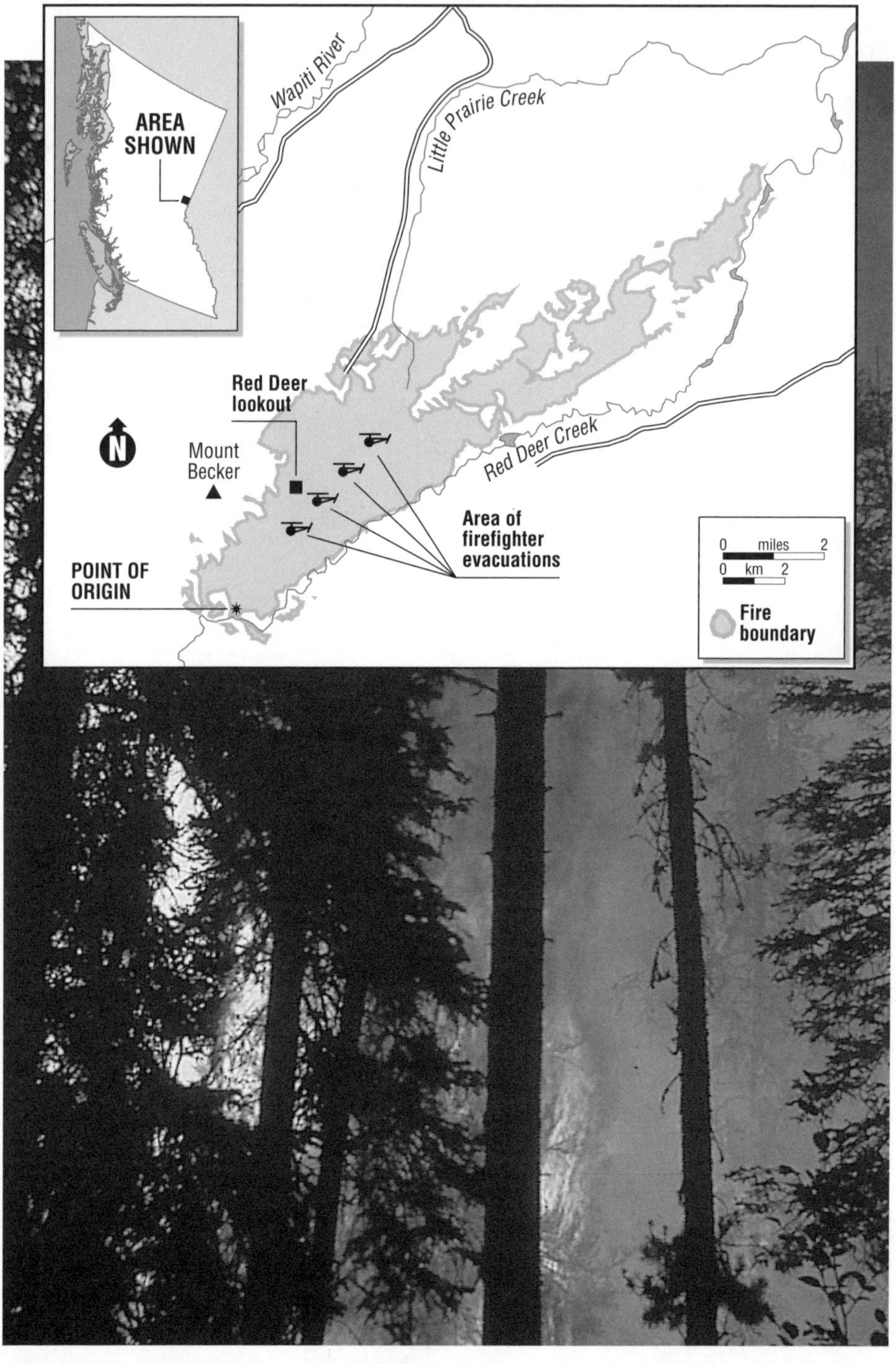

AREA
SHOWN

Wapiti River

Little Prairie Creek

Red Deer Creek

Red Deer
lookout

N

Mount
Becker

Area of
firefighter
evacuations

POINT OF
ORIGIN

| 0 | miles | 2 |
| 0 | km | 2 |

Fire
boundary

*A campfire abandoned at this unofficial campsite below Red Deer Falls sparked the 8,100-hectare Deer fire.* BC Forest Service photograph

In September 1987 someone carved out a campsite in the valley bottom of Red Deer Creek. The location selected was a thick stand of spruce about a half kilometre below Red Deer Falls, where the creek plunges over a 20-metre escarpment. Later, an Alberta pickup with two four-wheeled all-terrain vehicles in the back was reported having been parked at the site, though newspaper advertisements requesting further information on the campers drew a blank. Moose hunting season was underway, but Alberta hunters are allowed to hunt in BC only when accompanied by a professional guide. Whoever stayed at the site may have been hunting illegally or may simply have been exploring that ruggedly spectacular corner of the northern Rockies. Though their activities remain a mystery, their skills—or perhaps their principles—with regards to campfires do not. Abandoning a burning fire is inviting disaster into the country.

Big fall fires are not unusual in BC's north. Brian Pate once made a study of fire-related losses incurred by his employer, West Fraser Timber (now Chetwynd Forest Industries). Looking at a 10-year period, he found that most major losses occurred in September and October.

> Pate: By that time of year it's dried out. It's the end of summer, the trees have gone to sleep. When an evergreen gets ready for winter it builds up an antifreeze in the needles so they don't burst. And that antifreeze is very flammable. And all the grass that's been growing all summer is dried out. So basically you have spring conditions in the fall. And higher drought codes. So it's very hard to control. And then we get the winds. We get big winds here quite commonly. Anybody that ever fucked up a fire here would write, "Winds were extreme today. Winds were unexpected, up to 50." Even though someone like me would have said, "You'd better get out there and start bucketing on that smoke—you're going to have 50-mile-an-hour winds this afternoon."

Drought and high winds had established outstanding burning conditions at Red Deer Creek by the third week of September 1987. The abandoned campfire provided a source of ignition that smouldered for an unknown period before making its leap to infamy. And therein lies one of the central controversies surrounding the Deer fire. The Dawson Creek Forest District maintained a secondary lookout station with a view into the area where the impromptu camp was located, but at the time did not have a person staffing it. The question is, would a pair of eyes in that lookout have averted the impending blow-up? Pate believes it likely would have.

> Pate: There was a lookout that could have seen down there. It was supposed to be manned—a 54-day fire hire, they used to call them. The Forest Service was allowed to hire someone for 54 days, then they had to lay them off. It was a way of saving money. So we got into extreme hazard in September and they laid everybody off.

Pate was in the Red Deer drainage on September 21, supervising West Fraser's pre-logging preparations. Though parts of the valley had been previously swept by fire, the area West Fraser intended to log hadn't been burned for 140 to 180 years—an unusually long fire-free period for that region—and it contained exceptional stands of white spruce. Some staff from the Dawson Creek Forest District were also in the valley, and Pate discussed with them the extreme fire conditions that had developed, emphasizing that it would be prudent to have someone up in the Red Deer lookout.

The person responsible for fire suppression for the Dawson Creek Forest District was resource officer Paul Gevatkoff, who is now the district's operations manager. He maintained that staffing Red Deer lookout may not have prevented the Deer fire and that not staffing the lookout may well have saved a life. By way of justifying his first contention, Gevatkoff pointed out that a retired Dawson Creek forest ranger flew over the Deer fire's point of origin on the morning of September 22, perhaps within minutes of its transformation from smouldering campfire to raging inferno. The ex-ranger, on a pleasure flight in his own plane, circled Red Deer Falls for the benefit of a passenger. The pair noticed nothing in the way of smoke emanating from the unbroken spruce canopy near the falls. Though the campfire was likely spreading by this

*The light area near the photo's centre marks the location of Red Deer lookout, destroyed when flames roared up the hillside at right. Opinions differ on whether the Deer fire could have been prevented if the lookout had been manned.*
BC Forest Service collection

time, Gevatkoff said, winds could have dissipated smoke to the point that nothing showed above the dense canopy until the blow-up began.

Pate disagreed. A lookout person, he said, would have been scanning the valley well before the pair flew over Red Deer Falls. Early morning is the time to see smoke, while it's cold on the valley floor and smoke rises through it, before the wind starts and while the sidelight is just right.

At 1:30 p.m. Gevatkoff got word that a logger working in the adjacent Wapiti River drainage could see smoke rising above the ridge separating him from Red Deer Creek. Gevatkoff put in a request for air tankers from the Prince George fire centre, then headed south by helicopter. He and pilot Mike Maylin arrived at Red Deer Creek 90 minutes after receiving the logger's report. Finding the fire was no problem whatsoever.

> Gevatkoff: At that point the smoke was going to about 25,000 feet. The fire was indescribable, it was so volatile. It was burning up through the smoke. It was kind of a pulsating black and red mass. I've been with the Forest Service for 31 years and I've never seen a fire show that kind of intensity. It's awe-inspiring, and it's scary as hell.

By the time Maylin and Gevatkoff arrived over the valley, an air-attack officer flying ahead of the Prince George-based air tankers had already taken one look at that pulsating conflagration and—as had occurred on the Eg and Swiss fires—ordered his bomber group home. Flames had already destroyed the Red Deer lookout, located approximately seven kilometres northeast of the fire's starting point.

In Gevatkoff's estimation, roughly two hours would have passed between the time a person in the Red Deer lookout would have spotted the fire and the time the lookout was burned over. Given that the hypothetical fire spotter would have spent some of that time making the decision to run for it, he then would have been left, in Gevatkoff's scenario, with at best an hour's leeway before the fire arrived—perhaps not enough time to call for a helicopter evacuation or to escape on foot.

"That's sort of bullshit," said Pate. He argued that a guide-outfitter had a hunting party on the ridge where Red Deer lookout is located. The guide saw the blow-up begin and had time to get his party back to camp, pack up and ride out on horses ahead of the fire.

Responded Gevatkoff: "I guess it doesn't cost people a dime to say things."

The day the Deer made its big run, helicopter pilot Harvey Evans was on a ridge about 120 kilometres distant, working with a group of Shell Oil surveyors. The fire was well on its way to covering the 6,000 hectares it would burn by nightfall, by which time a snowfall of ash would begin descending on Grande Prairie, Alberta, 150 kilometres to the east.

> Evans: It was a beautiful, clear day, and you could see this mushroom cloud going up. I said to the surveyors, "I probably won't be here in the morning. I'll be on that fire, just as sure as hell." We got home that night just before dark, and the word was already in: they wanted the helicopter just before daylight.
>
> The fire came through a little gut at the end of the valley, so the wind was working like a venturi, and the thing just exploded down that valley. We had been working there the previous month, laying out a five-year cutblock for West Fraser. We'd just finished 10 days before. So their roads were laid out, their ribbons were in the bush, and it was all gone.
>
> I flew out there and met Brian Pate. He'd been out all night, leading Cats in the dark. He wasn't quite sure where the hell they'd got to. When I arrived there at daylight I got him on the radio and picked him up from the side of the mountain, and we started looking at the fire. When we got there it was probably 12 miles long.

Pate and his boss at West Fraser, Bill Rand—like Pate, a former Forest Service employee—began working the fire.

> Pate: We put guards around it and we were on it for three or four days. The fire had already burned up half of our cut block. We were building a road in there, getting ready for that winter's logging. We were building roads inside the fire because it was so big—12 miles, 13 miles long. We got the

fire under control and then the Forest Service took it over. It went along pretty quietly for about ten days. After about ten days you could just smell it in the air. But it was hot and ready to go. It just needed a little wind.

The Deer fire camp was established in the Wapiti River Valley. Pate drove there one October morning to discuss his concerns about the fire with

## — *Helicopter Rescues* —

While the evacuation of 70 firefighters from the Deer blow-up was the largest helicopter-assisted retreat in BC firefighting history, it is only one of many times that rotary-wing aircraft, as they're known in the trade, have played key roles in saving firefighters' lives. All such rescues come about when fires blow out of control, cutting off any prospect of escape on foot.

BC's second largest fire-related helicopter rescue took place on July 29, 1960, at one of many trouble spots scorching the province's southern Interior. The Mil fire was named for its point of origin at Mildred Lake near Lac Le Jeune. It was discovered by the crew of a converted American B-17 Flying Fortress imported by the BC Forest Service with the hope that its substantial water-bombing clout might help quell the overwhelming regional outbreak.

The Mil seemed to be a jinxed fire from the outset. Ten hours were lost getting firefighters on-scene because the B-17 crew inaccurately reported the fire's location. Then a communication foul-up resulted in Cats running short of fuel. Hours later, two Avenger air tankers inadvertently extinguished fires lit to burn off fuels inside a fireguard. Crews ran out of food. Wild, erratic flames forced firefighters to move camp twice in four hours. The B-17 crashed while landing at Kamloops' Fulton Field. None of its crew was injured, but the plane was out of service for the rest of the campaign.

One week into the fight, the Mil blew wild, trapping 41 firefighters and catskinners behind crowning flames. While the trapped men attempted to

Gevatkoff. "We ended up arguing, screaming at each other," Pate recalled. "I was saying that the fire was going to take off and they were going to lose people if they weren't careful."

One of the fire's problem areas was a south-facing slope above Red Deer Creek. The hillside contained a large amount of what fire people refer to as "unburned fuel"—trees that had escaped the initial blow-up. The situation was volatile enough that Harvey Evans had refused to land firefighters there.

get out of the fire area, assistant ranger Art Hill began walking into it, his route selected and communicated to him over two-way radio by a Beaver spotter plane circling above. As Hill reached the men, a wind shift drove flames across the area he had just traversed, eliminating any possibility of escape in that direction. The spotter then suggested the group try a retreat to the north, but their attempt in that direction was thwarted by intense heat. Advised to build a heliport, several of the trapped catskinners set to work. One of the Cats ran out of fuel immediately. The others ran low. Only a precisely placed parachute drop of diesel allowed the job to be completed. Another airdrop supplied the parched crew with much-needed food and drinking water. Some of the trapped men had by then been working for more than 24 hours, having been ordered to stay on fire lines straight through from the previous morning in an effort to complete desperately needed control guards.

By now Premier W.A.C. Bennett had requested assistance from the Royal Canadian Air Force. The RCAF's big twin-rotor chopper made its first rescue attempt at three o'clock that afternoon but was driven back by air turbulence created by the extreme fire activity. Five hours later Flight Lieutenant Pat Matthews threaded his craft down onto the makeshift heliport. Making four flights, he and his crew removed the last of the exhausted firefighters shortly before midnight. For the crew that had worked the previous day and night it was the end of a 41-hour shift.

"The men that were trapped behaved very well," Hill wrote of the ordeal, "with no evidence of panic or disorder despite terrific heat, smoke and danger of falling snags."

*Helicopter pilot Harvey Evans refused to land firefighters on the ridge where crews were nearly fatally trapped by the Deer blow-up: "I told them for three days not to put men on that ridge."* Keith Keller photograph

Evans: I just told Forestry, "No." I wasn't going to put crews in there. I just wouldn't take the bullshit. If they were putting crews out where I didn't think they were safe I just wouldn't put 'em in there. I felt responsible. Every time I voiced my opinion I got sent home. Two days later they'd phone me and I'd go back to work again. I got fired four times from that fire. It was on a jack-pine ridge drier than a popcorn fart. And it was going to go. It just needed the right combination and it was going to go. And you couldn't stop it. I told them for three days not to put men on that ridge.

Gevatkoff and his line boss were doing a helicopter assessment of the problem area when 80–90-kilometre-per-hour winds began to aggressively fan flames. Pilot Alan MacHardy dropped the pair off on a gravel bar, attached a water-bombing bucket to his big Bell 204 then, dipping water from Red Deer Creek, began attacking the trouble spot.

It became almost immediately clear that the fire was blowing out of control—clear, that is, to observers from the valley below. However, not all of the 70 firefighters stretched along the four kilometres of hillside recognized as quickly that they should be shifting their objective from suppression to evacuation. Two crews from Moberly Lake and Chetwynd were in particular danger from the escaping blaze. Among them was Harvey Evans' son. Another was Richard Knott, born of Métis heritage at Dawson Creek and at the time living in Chetwynd. Having begun fighting fire at 16, he was a veteran of eight seasons when called out to Red Deer Creek.

Weather had been causing a high turnover among fire crews and crew foremen. Nights at that altitude were getting frosty. The firefighters were sleeping in floorless tents, and many awoke mornings to find their blankets and, in some cases, their sweat-soaked clothes of the previous day frozen to the ground. They then stepped outside into fresh snow. Knott recalled that many people deserted the camp by piling into any Dawson Creek-bound vehicle. By afternoon, temperatures often climbed well into the 20s.

> Knott: It was really hot the afternoon the fire blew up. We were about halfway up the mountain, putting out hot spots on the edge of the fire. When I looked into the other valley, I could see a big mushroom cloud. I was thinking that the fire was hot over there, but nobody's saying nothing so it must be burning up that other valley. I thought we must be out of harm's way 'cause nobody's saying nothing to us. I'd been on lots of fires before and I didn't think nothin' of it.
>
> But this Hiroshima-looking cloud kept getting bigger and bigger. Then I looked up the hill and the guy that was working beside me had dropped his hose and taken off up the hill. At first I thought he was just going to talk to somebody, but I guess the foreman of the crew I was on got the call that we were to evacuate. He got everyone else on the crew together, but he didn't get me. So here I am with this little Econo hose, putting out hot spots, and this mushroom cloud is getting bigger and bigger.
>
> Danny Gladue was the foreman of the crew below us, and Danny's crew came running up the Cat guard. They were screaming, "Fire!" Screaming and hollering, "We've got to go! We've got to get out of here! We've got to go up the hill to the helipad!"
>
> I dropped my hose and ran out to the Cat guard. I ran around the corner and saw what I think would be called a rank six fire. [Wildfire behaviour is ranked on a scale from

one to six, with six being the most volatile.] It was rolling on the crowns of the trees. It was about 100 metres up the hill, barrelling through the trees. The noise was deafening.

I turned around and said, "The fire's right there, boys." Then I said, "Do you guys believe in prayer?"

They said, "Anything, anything." They were real nervous.

I started praying out loud. Praying and praying. We had to go towards the fire to get where we needed to go.

Some of them guys needed a good slap. They were to the point that they were near panic. One of the younger guys was screaming, "Every man for himself! Every man for himself!" He was just causing everyone else to panic.

I was with these young guys, running along the Cat guard. And then a thought came to me: my father-in-law, Darryl Street, was on Danny's crew. I knew I couldn't leave my father-in-law behind. I turned and went back down the hill to check on him. I seen Darryl. He said he was all right, but he said, "There's two more behind me. Check on them."

I went down to these two guys. One of them, the crew boss, was a pretty heavy guy. I let him go on up, and I went to the last guy. His name was Wayne.

Wayne Foster, born in the Lake Simcoe region of Ontario, had left home at 15, eventually making his way to BC's Peace River country. He had been working as a camp attendant in the oil patch for 16 years. While unemployed in the fall of 1987, he heard that the Deer fire was almost out and that people were needed for mop-up, so 42-year-old Foster walked into the Dawson Creek Forest Service office one day to ask about work. The next morning he was on a Greyhound heading for Red Deer Creek. It was his first fire.

Foster had been assigned to a chainsaw crew, his job being to throw bucked-up lengths of trees away from the fireguard. As he later noted, working as a camp attendant hadn't exactly kept him in good physical shape. He had worked for about four hours when, around noon, his crew broke for lunch.

Foster: We were eating our sandwiches when the bush started filling up with smoke. All of a sudden guys ran past us yelling, "Get out of here!" We had a foreman with a radio. If anything happened they were supposed to tell him from the air, but when the fire hit, everybody just dropped their chain saws and everything and took off. Every man for himself.

There was one guy quite a bit older than me, but he was in quite good shape, so he was able to go up the hill okay. To me it was like a mountain. I made it halfway up and the smoke got me. I'd say about another five minutes and I'd have passed out. I could hear this weird noise behind me. It was the fire in the tops of the trees.

I was on my hands and knees. I was just about ready to lay down. I was so scared. I was dry-heaving and coughing and I just couldn't make it. This guy come out of nowhere and grabbed me by the belt and just drug me up. He almost dragged me to the helicopter.

Knott: Wayne and I started going up the hill to the helipad, but Wayne just sat down. He was going to stay there. He said, "Just leave me here. I can't make it."

I looked at him. I said, "No." I said, "Wayne, I'm not going to leave you. We're not going to stay here. If you don't want to get up, I'll pick you up, I'll put you on my shoulder, and I'll carry you. You're a big fella, and I won't get very far. But we're not staying here."

Back then I was about six-one, 185 pounds. Wayne was about six feet tall, 240 pounds.

Wayne said, "No." So I grabbed his hand and basically pulled him up the first hill—the one I'd already run up and back down. There was fire all around us, on both sides of the guard. It was burning and the smoke was really low. Wayne started throwing up on me. I got Wayne and myself as close to the ground as we could, 'cause there was a foot and a half, two feet, near the ground where there wasn't a lot of smoke. If we were going to get any oxygen that's where we were going to get it.

So we were on the ground, breathing. I took off my coat and gave it to Wayne and told him, "Hold it over your mouth. Let's go." I held him by his hand and just kind of pulled him along. When we got to some of the hills where there was less smoke, I grabbed the coat. He would hold onto one end of the coat and I'd hold the other end and try to pull him along.

When me and Wayne were walking up the hill—while there was fire on both sides of the Cat guard—I prayed, asking for God's protection. I said, "Do you believe in prayer, Wayne?" I can't remember his exact words. I think he said, "I'll believe in anything right now."

Foster: I think I remember saying something about God.

*Richard Knott (back row, third from left) with other members of the Dawson Creek Wildcats, a Métis firefighting crew formed in the wake of the Deer fire: "I started praying out loud. Praying and praying."*
Photograph courtesy Dale Mineault

Knott: When we got to the first helipad, where the crews had gathered, the chopper pilots radioed down that they couldn't see the landing, the helipad.

Foster: The place where we were supposed to go for the helicopter, the helipad, was on fire.

Knott: So all the guys took off ahead of us again. I stayed with Wayne. I prayed with him. I remember that I kept telling him that I wouldn't leave him. I said, "Don't worry, Wayne. I'm going to stay with you all the time. I'll be here."

Wayne and I managed to make it to the next helipad. I couldn't really tell you how far it was, maybe half a mile up the hill—they were helipads eight and nine, that's all I know. When we got there, Wayne kind of collapsed from exhaustion. The bigger choppers came in, but the pilots couldn't see the landing at that helipad either through the smoke. *We* could barely see through the smoke. In my opinion, there was even more smoke at that helipad than at the first one we got to.

One of the crew bosses radioed in the first chopper. It came down through the smoke, landed, loaded 12 people and lifted off. And I swear that when the first chopper lifted off, his rotor blades were within 15 feet of the other chopper's skids. You could see them both right there. I wish I'd had a camera.

Had Knott been taking photographs he might have recorded the co-ordinated actions of Alan MacHardy and fellow helicopter pilot John Kennedy—two of at least three pilots who evacuated crews trapped across that flaming hillside. MacHardy had participated in several fire-line evacuations during his four decades of flying, and he would later have his own close brush with death on the Deer fire when a metre-long section of his main rotor broke off. (The incident, which he described as "one that you'd normally expect to be fatal," forced him into a barely controlled emergency landing. His chopper was

*A portion of the Deer fire, with smoke still rising from some of the blackened areas. Strong winds were responsible for the fire's long, narrow profile. The outline of previous fires of similar proportions can be seen in the regenerating forest*

badly damaged in the accident—not by the landing but by the violent shaking it underwent as MacHardy wrestled it to the ground.)

MacHardy called the Deer "just a normal fire that went berserk." Smoke around the helipad was thick enough that Kennedy could see almost nothing around him as he prepared to lift off with his load of firefighters. Though MacHardy directed Kennedy as best he could from his holding position above, Kennedy had to switch from visual to instrument flying to get off the hillside. As Kennedy departed, MacHardy noticed another batch of firefighters arriving at the pad. He felt his way in through the smoke, loaded up and evacuated the group that Knott and Foster had joined.

> Knott: As soon as the first chopper lifted off, the other one was down. Twelve of us got on the second chopper—there was only 12 seats. There was one guy left. He said he was going to stay behind.
>
> The pilot said, "What'll he do?"
>
> We said, "There's only 12 seats in here."
>
> The pilot said, "Tell him to get in. I'm not comin' back."

*in the lower part of the photograph. Richard Knott and Wayne Foster were evacuated when the Deer blew out of control.* BC Forest Service collection

The last guy's name was Don Nelson. He came in and laid on the floor. The pilot lifted off, took us down the valley to the creek. After we landed, I left everyone to go and be by myself. And then the emotions that I wouldn't let Wayne see hit me all at once. I just went off by myself and wept. That's how it ended.

Foster: When they dropped us off on the bank of the river, they counted us to make sure we were all there. I went down to the creek, took my shirt off, soaked it in the creek and put it over my head. I was trembling. I can think of better ways of dying than getting burned alive.

They wanted to give me oxygen. I was so mad I just said, "No, I just want to go home." The boss—I forget his name—said, "You hang on until tomorrow and we'll take you home." So I did. I wouldn't even eat supper that night I was so mad. Mad and scared. That night I was standing around the campfire and a lot of guys come up to me and said, "Are you okay?" And I said, "No, I'm not okay. I'm mad."

They took about 20 or 30 of us out the next day. Not everybody left. But this was my first day on the job. It was my first fire—and my last, too.

Brian Pate: When the fire took off, I went over and helped organize the evacuation. After supper we had a meeting about what was going to happen the next day. It turned into kind of a debriefing about what had happened. Everybody said their two bits' worth. And then they all left. There was such shock in the Forest Service that they all disappeared.

There was an Atco trailer and there was a bunch of guys that stayed in there. I was sitting in the trailer, with 150 men outside, 11 helicopters—and nobody from the Forest Service. So I got on the radio and said, "Okay, I need another 6,000 gallons of fuel . . . " This and that. They'd de-manned a bit but I got all the equipment back. I phoned the regional office, got the Protection guy in charge to come up the next day and debrief me. What happened was, they put Bill Rand back in charge of the fire again and put me in as line boss.

We got the fire back under control again, turned it back over to the Forest Service. They brought in two other guys—good guys, but they didn't know the area that well. We're into October by now, and it was getting down to minus 15 at night, so all the hoses and pumps had to be drained each night. Then, during the day, it was 25, 26 degrees above. When it freezes it makes it worse 'cause all the moisture from the ground gets drawn to the surface. Then in the morning, when the sun comes up, the moisture blows away. We started visiting these guys they'd brought in, saying, "You'd better watch out, this fire's going to take off again."

"Oh, no bloody way." This kind of talk. They'd had a bad day but they'd held the fire.

We said, "No, the fire's going to take off tonight."

And it did. Temperature inversions happen a lot here.

# — *Happy to be There* —

George Kalischuk began fighting fires in Alberta in 1956, eventually becoming a ranger in the Grande Prairie area. He later joined the BC Forest Service and worked around much of the province's north before retiring in 1987. He was stationed out of Chetwynd in 1985 when the Rum fire began in the heavily timbered Sukunka River Valley.

Kalischuk: The Rum fire was the same year as the Ran fire. Dawson Creek was involved in the Ran fire, and it just kept ranning and ranning and ranning. We had the Rum fire and the office initially sent out a standby crew. This was in the mountains towards Hook Lake in rough terrain. I could tell by listening to the radio that these guys weren't doing too well—"Oh, now the fire's candling!" This went from early in the morning till almost noon. I thought, I'm going to end up on that one once it gets out of control.

Sure enough they called me up—they were intelligent enough to do that. But I'm swearing away to myself because they should have done that right off the bat. When I got there the fire was starting to go up this hill. And all these guys were doing was cutting a heliport. I guess they cut two heliports, but they hadn't done bugger-all with the fire.

I landed there and suddenly the fire started coming up the hill. It had about 150 feet to go before it reached us. And there was one too many guys for the helicopter. The pilot said, "Jesus Christ, let's get the hell out of here." The guys jumped in. There wasn't a seat left for me, so I said, "I'll wait. You come back for me."

The pilot said, "Bullshit. Jump on the skid." So I jumped on the skid and he took the thing up. I guess he flew a half-mile or a mile with me hanging onto the skid before he landed. I was happy to be there.

You get 15 degrees right on the surface, and cold air above and below. And the wind blows all night. That's what happened. That was the last run the fire took. We were back on the fire again, called out to control it a third time.

In retrospect, Paul Gevatkoff acknowledged, with Red Deer Creek as an ample water source, more people could have been assigned to dousing the area where the blow-up occurred.

Gevatkoff: What we should have done is put more people in that unburnt fuel, sped up the suppression process, so we wouldn't have had fire there after 11 days. In hindsight we would have acted differently. We had two crews in there, but I guess we underestimated how much fire there was.

A counsellor was made available to Forest Service staff traumatized by the near-disaster. The counsellor could also have worked with firefighters caught in the fire's rampage, Gevatkoff said, but they would have had to ask for the opportunity. None did.

Foster: I went to Forestry in Dawson Creek later and they gave me $90 for the four hours work I did. Which I accepted. But I was really, really mad. The secretary asked me if I wanted to talk to somebody about what had happened. I said, "I don't want to talk to anybody. I'm just lucky to be alive." I said, "I'll never fight another forest fire again. I'll go to jail first." I told the Forestry that Richard saved me, but he never got acknowledged for that, which is a shame. If people get in trouble, their names are on the news. If somebody saves someone's life, why isn't *their* name on the news?

I've seen Richard several times since then and I shake his hand every time I meet him. I've thanked him, but that's all I could do. I don't have much money. But if it wasn't for Richard I wouldn't be here, I'm convinced of that. If it wasn't for Richard I think I would have been burned alive.

# CHAPTER 15

# Wayne Goes to Machu Picchu

The BC Forest Service has developed such an international reputation for excellence in wildland firefighting that dozens of nations from as far away as Africa have sent delegations to learn what makes this province's suppression organization so successful. Similarly, BC firefighters have been called on to help with particularly troublesome fires in Malaysia and the Galapagos Islands, as well as elsewhere in North America. Wayne Langlois's turn came in 1992 when he travelled to Peru, where seasonal agricultural burning had gone out of control near the ancient Incan city of Machu Picchu. Langlois was accompanied by Peter Miller, who was then his overhead team line boss and now works in private industry.

Langlois: I was at Salmon Arm as an overhead team boss, and I got a phone call in the office. But it was in Spanish, and nobody could understand what was going on. I'd had a call from Victoria, saying, "They're having some problems in Peru. They want to talk to a fire expert, somebody who knows something about large fires." I had said, "Fine." I'd help them as much as I could.

So the phone call came and it was in Spanish. I finally rounded up a guy in the office that could speak just a *poquito*—a little bit of Spanish. He got a ways with it: there was a large fire in Peru, at Machu Picchu, and could I tell them over the phone what they should do with it? I said, "Tell them as best as you can that they should request that I go down there." 'Cause there's no way I could help them over the phone.

So I guess they went through the channels, the Canadian International Development Agency in Ottawa and so on, and three or four days later I got a call saying they wanted me to go to Peru to help with this fire.

We arrived in the Lima airport. They'd spent three days getting an interpreter for me. They got him out of Iquitos on the Amazon where he was working on a project for CIDA. And my name's Langlois, a French name. He spoke French. And Spanish. I don't speak either of them. But he did know enough English that he came along with us and did very well.

At the airport we were met by a general, a man of probably 55, 60. And he's totally dead drunk. Totally. They took us into a room that they'd set up at the airport. They had some pictures and maps. And this general, who's dead drunk, is going to explain about the fire, what's going on. So he's standing at the front of the room. And he keeps looking at me—'cause I'm the fire boss or whatever—and he keeps saying, "Wicky. Wicky." At first I had no idea what he wanted. But he's saying, "Whiskey, do you want some whiskey?" What a sight. And I'd just been travelling for 42 hours. That was my introduction to Peru.

Some of the characters were out of this world. They lined us up with a contact from France. They gave him to us as a contact because they were sending over a couple of people from France, a major and a colonel, to help us. Which was the biggest hindrance in the world 'cause they'd never seen a fire before. I don't know to this day why they were there. I didn't understand the politics. All they did was party and chase women.

Anyways, they gave us this fellow as a contact. I'm not sure to this day what he was doing in Lima, but he was from France, and he had something to do with oil wells and rigs—sort of an expediter. But he had a lot of enemies, as I found out later. He was always looking over his shoulder, and he wouldn't drive here and he wouldn't drive there. He had a little white Volkswagen, and he drove us out to the airport so we could catch the plane to Cuzco. When we got back from the fire and went to his car, it was full of bullet holes. He wasn't surprised—this had happened before. And he was leaving the country.

Everywhere we went in Peru, when we went out for supper or whatever, there were bodyguards. If we went into a little café, there'd be a guy with a machine-gun. We thought it was kind of fun at first. Then we thought, Hey, there's got to be a reason for this.

We flew out of Lima the next day to Cuzco. I thought the deal was that they would leave me in Lima for a day or two to acclimatize. I'd left from Salmon Arm, flown to Vancouver, Miami and landed in Lima. And the next morning they flew me out. And when you go to Cuzco you're stepping up to around 14,000 feet. Later I got up to around 15,000, 16,000 feet, and got pretty lightheaded—altitude headaches. It took a lot of getting used to.

I think it took another two days to get to where the fire was, at Machu Picchu. They didn't have any fuel for their helicopters. They gave us an old 212 military helicopter with a big machine gun mounted in the back of it. But no

*Wayne Langlois surveys the Incan city of Machu Picchu, Peru. He was one of two BC Forest Service staffers sent to fight fires started by farmers in 1992.* Photograph courtesy Wayne Langlois

fuel. It took two days before they had fuel to fly us out to the fire and have a look at it.

The fire was started by farmers. They do it traditionally every year, burning off the hillsides to create feed for their cattle, goats, llamas, whatever. It's no problem in a normal year, but it just happened that it had been a real dry year, and the fire moved up along a railway line, then up toward Machu Picchu. Being one of the world's tourist meccas, it closed off the tourists so they couldn't get through. And that was their big concern: Can you turn the smoke off? Get the tourists back?

Eventually we got enough fuel to leave Cuzco. We got about halfway to the fire and the pilot points to the fuel gauge. He says, "We can't make it to the fire." He put it down in a little village, in a schoolyard, and we had to wait there for another day to get to the fire.

We eventually got to the fire, and it was 60 kilometres long, right from the railway tracks in the valley to the tops of the mountains. We never did see all of it. And they didn't have any resources to fight it. Nothing. We came up with four different plans for them. But they didn't have the money to do any of the aerial stuff. I tried to get the Americans. They had lots of helicopters in Panama, so I got ahold of American people through their Washington system, asked if they'd give us a couple of high elevation helicopters to work with. But no—"We're not going to help that country, no way."

So we're in a helicopter that'll only fly to 12,000 feet, with a fire up to 23,000 feet. And nobody's going to help them. So we had to make up a bunch of plans that didn't involve a lot of money or equipment, like picking a valley and putting in a hand line or laying down some retardant with helicopters or whatever. The fourth plan was to leave it alone and wait for the rains to put it out. That's actually what happened. There really was no problem for the farmers—it was probably doing them a lot of good—but it was certainly shutting off the tourists. It probably got to within three or four kilometres of Machu Picchu. It's all rock, but in our recommendations we felt they should do a lot of environmental work after the fire, which they never did—set up a prevention type program with the local *campesinos* and farmers to educate them about the best time of the year to burn. 'Cause you're not going to stop them. They've been doing it since time immemorial. And since then they've had a fire burn over there, and they're having a lot of problems, like soil erosion.

One of the helicopter pilots we flew with spoke very good English—he'd been trained in Texas. He understood what we were doing—what we were trying to do—and he just shook his head and said, "No, that'll never work here." I could see that for myself after a day or two. He wrote me a letter a month or so later and said, "Just about five days, six days after you left the rains came and put everything out."

One thing that really got my attention, other than the machine-gun poking out the back of the 212 helicopter, was that when we were checking the fire with the helicopter, they never really flew *up* the valleys. They'd sort of come in at a cross angle and look, then go up higher and look. And I find out that the cocaine dealers and people growing all the coca leaves have cables stretched across the valleys to catch the narc agents and military and so on. They have small cables strung across creek draws and valleys. You fly up there in a helicopter and it catches you and takes you right out of the air. That's why they'd come in at the side and take a look. I found that out about two days after I started flying there.

We had a meeting with their military people and their fire people one night. I'd just come back from a day on the fire where they were cutting some guard—I was trying to stop the fire in one area, keep it from jumping a creek. Back at the meeting I could see that everyone was quite excited.

But they were all speaking Spanish. I said to my interpreter, "Luc, what are they talking about?"

He said, "The snakes."

The fire had chased all the snakes out of the fire areas and they had congregated along the edges, in the green stuff. And there'd been three or four people bitten that day. From that point on I didn't walk in anything that wasn't black.

Near the tail end of it we had to write up a report, our four recommendations. I wrote up the report, got it all typed up. I was supposed to present it back in Lima. The French wanted a copy of it first so they could read it. What they did, they took that report and they went off to Lima a day ahead of us. And they went to the press with it as if it was their own. It came out in a newspaper in Lima: the French have four different plans to fight this fire. The Canadian ambassador I was working with was mad. Man, was he mad! He was mad at me, too, for giving them the report, but these guys were supposed to be part of my team. He said, "No, they do this sort of thing all the time." I wasn't aware of those kinds of political games.

So they called a press conference for the next day. I still remember that, sitting down in the basement of the Canadian embassy. A lady there said, "Oh, we hold press conferences all the time. Nobody ever shows up." I'll be damned if at 10 o'clock, when it was scheduled to start, they started coming in. There was three TV stations, radio stations, seven or eight newspapers—about 50 people. So the story finally got out: it was Canada over there trying to help them.

On the way back they detained me in Miami. Again this was getting into politics that I didn't understand that well. Those American customs agents didn't care if I was fighting fire or whatever. They looked at my passport and I've got two Cuban stamps 'cause I'd been to Cuba a couple of times on holidays. And now I'm just coming out of Peru.

They questioned me for an hour, an hour and a half. "What's the connection between Peru and Cuba? Why are you going to countries like Cuba?" I guess they figured I was in the drug business or something. That ended the trip.

# CHAPTER 16

# Ash Wednesday

"It looked alive. It looked like a huge serpent. I've never seen anything that awesome in all my life."

—(Salmon Arm resident describing the August 5, 1998, Silver Creek firestorm)

There are times when weather ceases to be a convenient conversation filler and becomes everything that counts. Wind, in particular, has a way of focussing firefighters' attention like nothing else can. The story of the 1998 Silver Creek fire can be seen as the story of three winds: one not forecast, one under-forecast, and one which, without materializing, motivated the greatest evacuation in BC history.

But before there was wind there was lightning—lots of lightning. And the lightning arrived after impeccable conditions had been created for its terrestrial spread in the form of wildfire. As a source of these conditions El Niño would be repeatedly named—the cause, it seemed, of everything naturally disastrous in 1998. El Niño had emerged with a vengeance the previous winter, with the result that Africa, South America and Southeast Asia had endured unusually extensive fire seasons.

# Ash Wednesday (Salmon Arm, 1998)

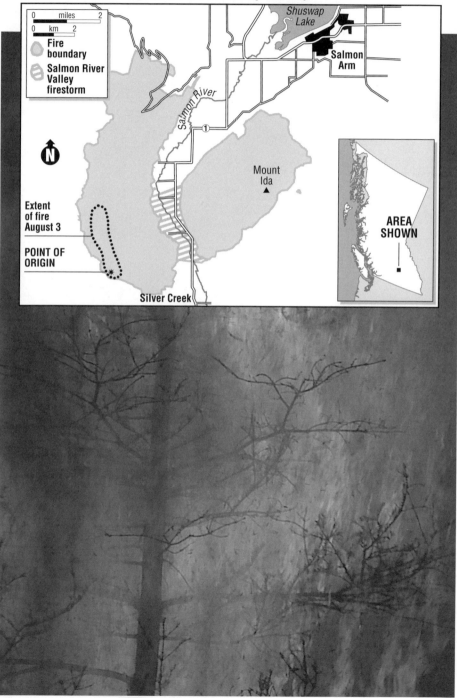

With this preview, BC's Protection program added more human and mechanical resources to its usual fighting force, a supplement they referred to as "the El Niño addition." Then came June, when a quarter-billion-dollar suppression effort couldn't prevent 1,600 wildfires from burning across 700,000 hectares of Alberta's forests and grasslands. Well aware of the devastation next door, since many of this province's firefighters spent June 1998 in Alberta, BC recruited a further "emergency complement," crews and resources from Alaska and several Canadian provinces.

By July, there were record high temperatures throughout BC's Interior, establishing unforgiving burning conditions. Fire managers monitor a range of complex indices which help them calculate the risk of fire ignition and spread. They also use a less scientific rule of thumb known as crossover, the point at which the temperature in Celsius exceeds relative humidity. A few days of crossover is a clear danger warning. Salmon Arm had hit the crossover point on July 23 and was still there nearly a week later when BC's network of remote lightning detectors signalled an intense concentration of strikes centred in the Okanagan–Shuswap. Given the extreme fire hazard, those detectors were in effect recording the number of lit matches dropped on a landscape primed to explode. Over the next 10 days, 606 new fires were reported, almost all of them caused by lightning.

One of those celestial bolts came to earth on the afternoon of July 29, striking high in the Fly Hills above the rural community of Silver Creek in the verdant Salmon River Valley, 20 kilometres south of Salmon Arm. That electric spark blossomed to life on a heavily timbered, tinder dry, nearly inaccessible south-facing slope. An air tanker en route to another fire was diverted to Silver Creek within the hour. Though aircraft dropped 85,000 litres of retardant that evening, K30285, aka the Silver Creek fire, had grown to 10 hectares and was burning vigorously when action ceased for the night.

Roy Benson, a fire zone manager based in Vernon, was airborne on the morning of July 30, 1998, attending to a number of lightning-inspired blazes in his north Okanagan jurisdiction. His helicopter pilot, having overflown Silver Creek on his way to Vernon from Salmon Arm, told Benson that the fire was "ugly" and that whoever the Kamloops Fire Centre delegated to fight it was going to have one hell of a time. The pilot had barely uttered these words when Benson got a call over his headset: he and his overhead team

*Fire boss Roy Benson.*
Keith Keller photograph

were to report to Silver Creek, pronto. In the air over the fire by noon, he agreed wholeheartedly with his pilot's grim appraisal.

Benson: When I first saw it, the fire was rolling down off the rock face into the north fork of Silver Creek. It was about 35 hectares on really steep ground with lots of fuel, and it was burning quite aggressively. There was a fair amount of blowdown in places—a lot of wind-thrown material on the ground. That's what was hampering some of the air tanker drops. The downed trees were preventing the retardant from getting to the ground, so the fire was burning under the windfalls and out the other side. It was so dry that single sparks would fly and land in downed material and that would be enough to get the fire up and going again.

It was this severe terrain that had previously prevented the company with timber rights to the area from building a logging road into the canyon where the fire was located. Later, motivated by extreme necessity, heavy equipment would be used to cut a trail into the fire-start area, allowing suppression crews to get in and cut guards by hand across slopes that Cats couldn't work—which included most of the places where guard needed to be built. In the meantime, Benson needed some form of immediate access to the canyon and, more importantly, an escape route for anyone having to get out in a hurry. He sent in a two-person Rapattack team—Chris Mayer and Steve Kidd—to build a helipad on a rocky knob jutting from the hillside near the fire's lower edge. The pair dropped in from a hovering helicopter at six o'clock on the evening of July 30.

*The rock knob that burned over shortly after two Rapattack firefighters were evacuated by helicopter. Fire crews later used this knob, located just below the fire's nearly inaccessible point of ignition, as a platform for portable water reservoirs (visible in photo).* BC Forest Service collection

Mayer: The first hour went pretty good. Then about an hour and a half into it the wind started picking up. The fire was burning fairly vigorously, but we were separated from it by a kind of draw. Then the winds caused the fire to spot over to our side, and it spotted into the slash we'd created from falling all the trees. That's what blew up on us.

We had a helicopter circling around, keeping an eye on the fire for us. It was ready to pick us up as soon as we finished our job. But the burning slash created so much smoke that it made it hard for the chopper crew to see what was going on. They sort of lost sight of us. We finally got the pad finished, but eventually even it was ablaze. There was a rock slide nearby, so we just gathered up our gear and chucked it down there, hoping it would survive. The helicopter was able to get down just long enough for us to jump in. The pilot was quite hesitant. The heat was intense, and he was worried about melting the bubble on his helicopter.

Salmon Arm Observer *photographer James Murray captured this award-winning photograph as flames tore up Mount Ida during the Ash Wednesday firestorm.* James Murray photograph, *Salmon Arm Observer*

The rock knob from which Mayer and Kidd were chased burned so cleanly that evening that it later served as a safe base of operation for fire crews pumping water out of portable helicopter-filled reservoirs. Over the following three days the Silver Creek fire grew slowly. By August 3 the ground-level suppression force included various heavy equipment, 66 firefighters and more than two dozen support people. Working the fire from the air, in addition to 15 helicopters and two retardant-dropping aircraft, were four heavyweights of the water-bombing world: a pair of Bombardier CL-415s from Quebec and the venerable *Phillipine* and *Hawaii* Martin Mars from their base at Sproat Lake.

Early on the afternoon of August 3 unpredicted winds caught firefighters by surprise, driving flames across control lines and igniting a spot fire in a heavy accumulation of downed trees. Helitankers working the main fire switched to the spot immediately but couldn't stop it racing across the parched hillside. When the blow-up subsided, a 300-hectare fire stretched in a cigar shape above the Salmon River Valley. With the ante now considerably raised and with fire boss Benson away supervising a second fire on Shuswap Lake, O-team operations chief Jim Mottishaw requested that the fire commissioner's office put Silver Creek residents on a 10-minute evacuation alert.

Bulldozers were sent to cut parallel control guards below the fire's lower edge, one as high up the precipitous slope as Cats could work, the other on more workable ground close to the valley floor. The lower guard would be used as the anchor point for a burn-off should Benson elect to torch the entire hillside rather than letting flames burn downhill on their own terms. But burning off from the valley floor didn't much appeal to him. As he saw it, even if catskinners succeeded in building the parallel guards, there would be little to prevent such a pre-emptive burn-off—which would have covered a minimum of 1,600 hectares—from escaping. Benson was also aware that, as one local observer commented, "Politically, it's not a very popular thing to intentionally burn a community's mountain." Burning southward across the hillside, an escaped burn-off would threaten rural Silver Creek. Escaping northward, flames would have free rein on a continuous forested slope running toward the town of Salmon Arm. A token containment line in the form of a rough Cat guard was started up the hill at right angles to the two parallel guards, but Benson had few illusions about its ability to box in rampaging flames. "If the fire challenged the guard," he said, "it was going to go over it—conditions were that volatile."

With the Fly Hills time bomb ticking ever louder, the Ministry of Forests convened a public meeting on the evening of August 4, their goal being to address residents' concerns and "to explain the nature of the event going on in the hills above them." It remains an open question as to how well each side conducted itself that evening or whether, under such trying circumstances, either side could have done better.

Benson, along with Judi Beck, a fire behaviour specialist, addressed 150 people in the Silver Creek community hall that evening. Beck recalled a "very negative" tone was quickly established. Rather than learning about the fire's potential, she said, residents preferred to criticize the Ministry of Forests' real or imagined handling of the emergency to that point. (Rumours in Silver Creek were rampant: the fire had been ignored rather than fought from the start; the fire had been used as a training exercise for Rapattack teams; the fire had been allowed to burn out the canyon it started in to eliminate a pine beetle infestation; air tankers had been used too little, too late.) Being a bureaucrat, a scientist and, perhaps worst of all, a representative of the government, Beck became a lightning rod for residents' discontent. "What could you do," she later commented, "when you have to stand up in front of the public and say, 'We're losing this battle. There are not enough resources in the whole world for us to put this puppy out'.?"

Many Silver Creek residents, however, felt they were not given a sufficiently explicit sense that K30285 was a restless puppy with an enormous bite. Some also felt that their opinions were not given due consideration. Riding instructor Wendy Hawes came away from the meeting with exactly that feeling.

> Hawes: One person asked Forestry if they were aware of the strong crosswinds that we could get in the valley at that time of year. There wasn't really a response to that. If they were so knowledgeable, why didn't they know about those winds? They sure could have asked us about them. These winds only last maybe a day, but they come at about that time of year, and when they come you want to have everything strapped down. But they avoided the question when it was brought up. It was bizarre.

An RCMP officer in the audience that night later told an interviewer that, having heard what Benson and Beck had to say, "*My* family would have been gone...I would have had them moved, no question about it." Rapattack crewman Chris Mayer also attended the meeting. Now a software design engineer with Microsoft Corporation in Seattle, Mayer remembered that the Ministry of Forests' message "stressed the unlikeliness of the fire changing direction and blowing up."

Beck and Benson had gone into the meeting after receiving a preliminary forecast for a cold front—a wind—arriving the following evening, but neither mentioned that forecast to the assembled residents. "On the August fourth evening we had no strong indication that there would be a wind event on the fifth," Benson later commented. Since it would not be confirmed until the morning of August 5, Beck added, there was no point in instilling unnecessary panic. There would still be time enough for appropriate warnings to be issued and acted upon, so she and Benson outlined the difficulties of fighting the existing fire. They played the wind card close to their chests and hoped for the best. What they got instead was Ash Wednesday.

*Tractors, trailers and pickups are loaded with household possessions as Silver Creek residents prepare to flee the oncoming flames.* James Murray photograph, *Salmon Arm Observer*

The morning forecast on Wednesday, August 5, called for the cold front to arrive around suppertime, packing westerly winds of 30 to 40 kilometres per hour. The forecast was wrong on three important counts. Though the wind was westerly, it struck just after noon. Gusts were estimated at more than 100 kilometres per hour. And though the wind was predicted to last approximately two hours, it blew for 12.

As air rushed over the Fly Hills, the interaction of wind and topography generated a downslope flow. With winds blowing at right angles to those of August 3, a very long fire with a narrow front suddenly became a shallow fire with a very long front. Swept along in an invisible river of downpouring wind, flames were magnified into an inferno that poured toward the lush, agricultural Salmon River Valley.

Scientists measure fire intensity in kilowatts per metre (kW/m). An electric baseboard heater generates approximately 10 kW/m. Skin exposed to nine kW/m will incur second-degree burns in nine seconds. At peak intensity the Silver Creek blow-up was generating an estimated 60,000–80,000 kW/m, Beck said, some of the most potent fire conditions ever measured, including those created by fires intentionally lit to study extreme fire behaviour.

With flames racing downhill at 80 metres per minute, the O-team's Jim Mottishaw called for an immediate evacuation of Silver Creek residents. Ash and smoke rose to form a roiling convective column that first towered above the fire and then, pushed by powerful winds, bent into a Stygian rainbow arching across the Salmon River Valley. Embers and chunks of burning wood, some the size of baseballs, rained from the nightmare cloud, the deluge igniting houses, barns, fields, and in one case, household possessions piled in the back of a fleeing pickup truck.

Chris Mayer was among the scores of emergency response personnel sent into the Salmon River Valley as the August 5 evacuation got underway. With the main fire now clearly beyond human intervention, firefighters were assigned to protect houses threatened by spot fires. Working from a foam-generating fire truck, Mayer and his three-person crew began laying out hoses in a rural subdivision. As the situation deteriorated, Mayer called the Salmon Arm Rapattack base for backup and was joined by a two-member crew consisting of first-year firefighter Lisa Scharf and team leader Mike Worobey, who had worked on rap crews for seven seasons and was about to pursue, on a Rhodes scholarship, a doctoral degree at Oxford University. (Worobey has

*Powerful winds drive fire down out of the Fly Hills and into the Salmon River Valley and the farming community of Silver Creek. Flames advancing at up to 90 metres per minute sent residents and firefighters scrambling for safety.*
Blair Borden photograph

since earned his PhD in evolutionary biology, receiving the additional honour of being named a Fellow of St. John's College, Oxford University.

The combined crews were tackling a spot fire in long grass near houses close to the Salmon River when the winds began to pick up. Conditions became increasingly volatile. Flames raced through the grass in one direction for a short distance then shifted 90 degrees and raced elsewhere. A barbed wire fence separated Mayer's crew from their portion of the fire. Noticing that his team members were about to climb the fence, Mayer reined them in.

> Mayer: I called them back because I didn't want them to get stuck on the other side of a barbed wire fence. I wanted them to come back and get cutters and remove the fence properly before they went in there. And in the 10 or 20 seconds that it took to get cutters the winds had swirled the fire around and jumped it to the other side of the road.

*Rapattack firefighter Chris Mayer. "It was like sticking your face into a blow dryer," he said of being caught in the Ash Wednesday blow-up.*
Photograph courtesy Chris Mayer

Nearby, Worobey had noticed that in an already hot and dry situation, conditions suddenly got "really, really hot, and really, really dry." At that moment the big wind hit. The grass fire flared and the sunny blue-sky day instantly transformed into a frenzy of flames and swirling black smoke. The rap teams realized simultaneously that their new priority was to save their own lives, but as he prepared to run, Worobey was bothered by the niggling concern that he just might be turning tail too quickly.

Worobey: I know that from talking to Chris afterwards, the feeling we had was that we weren't too sure, that maybe we're being a little bit overcautious, but that we should probably get out just to be on the safe side. I know that when it went pitch dark there were lots of other people around there that did the same sort of thing—dropped what they were doing.

I ran to the back of my truck and detached the hose and just left it there. I left everything there, including my equipment bag. I told Lisa to hop in the truck. We started driving but it was pitch dark. We just had to drive down this road that went parallel to the Salmon River until we got to the bridge. But it was *dark.* I put the headlights on out of reflex, but there was zero visibility. So I had to drive a little bit slowly, because I had to drive by feel, not by sight.

I thought for what seemed like a long time—probably about 40 seconds—that I might not see the light of day again. That's the first time that I'd experienced that in seven years of firefighting. There were flames and there were embers flying. It was kind of like being beside a jet engine spitting out bits of burning stuff. And it was really, really hot, like being in a sauna with a high wind.

It certainly seemed ironic that for the seven years that I'd been fighting fires, I'd always assured myself that the dangerous parts of the job were the parts that we couldn't control, like flying in helicopters and using a chainsaw to take down a burning snag. I thought that those were the dangerous parts, but that with a fire you have the option of not going in, that you have the ability to keep an eye out and predict what's going to happen—to anticipate. But in this situation that fire did things that I certainly

*Mike Worobey, who made the jump from firefighter to evolutionary biologist at Oxford University. "I thought for about 40 seconds that I might not see the light of day again."* Photograph courtesy the Worobey family

didn't anticipate. And my sense was that there were a lot of people who had been working on fires for 30 or 40 years and that it did things that they didn't anticipate as well.

After about 40 seconds the skies parted, and I could see light through the black. At first I wondered if that light was a big ball of fire. But it was a little bit too white, and it got whiter and whiter. And at some point I realized that it was going to be all right, that we'd just passed through the fire while the fire jumped from one side of the valley to the other.

But Chris and the other guys were still back there, and I didn't know what had happened to them. When I got to a place on the road where it was safe to stop, I let Lisa out. I wondered about those guys for a few seconds, then I started to turn the truck back around. In the confusion of the moment I started wondering whether, back when Chris was walking towards me, whether when he said, "We're out of here," and I turned around quickly and got in the truck, maybe he'd also wanted to say to me, "Our truck is stuck," or "We don't want to take our truck 'cause it's too slow," or "We can't turn our truck around and can we all pile in the back of your truck?" Or whatever. I'd just turned my truck around to go back in when I saw through the smoke this big red truck just chugging along. And out they came. I remember having my hand out in front of me with four fingers up, trying to see the driver and say, "Is there four of you? Is there four of you?" Sure enough, there was.

As Worobey soon learned, his intuition about Mayer and his team had been pretty much right on.

Mayer: You couldn't really stand in the winds when the front hit. Mike was nearby, and I remember turning to him. We both knew, and we both gave the call to evacuate. I was still a ways from the fire truck, and I ran back towards it. Because the truck was facing the wrong way, I was running back to tell my crew to forget the truck and just run for it. The way it was facing was where the bulk of the fire was, where the fire took the first houses. It had burned in a kind of horseshoe around us—a dead end. The road bed was built up above the surrounding land so you couldn't drive off the road to turn around. But Jeff Croft, one of my crew members, somehow got the truck turned around. The fire truck was between me and the fire, and I remember stepping out around the truck when I got near it, and it was like sticking your face into a blow dryer. Instantly my eyes crudded up with flying soot and stuff, blocking my vision. But the truck was turned around, so I jumped in and told Jeff, "Okay, drive!"

He was like, "I can't see!"

I said, "Drive straight. If you hit something, start pushing it."

As the firefighters realized after getting to safety, fierce winds had stretched the ember-laden smoke cloud across the valley to the heavily forested foot of Mount Ida, the area's predominant landmark. From innumerable spot fires a new flame front emerged and raced up the mountain's southern flank at speeds even higher than those attained when the wind-propelled fire had raged into the valley bottom. As flames surged toward Ida's peak, Wayne Langlois and helicopter pilot Guy Denys noticed a car parked at a mountainside trailhead. Langlois had retired from the Forest Service that spring but had been called out to help co-ordinate helicopter operations on the fire. Later, after being evacuated from his own Salmon Arm home, he would comment that, "I've never been beaten up this bad by a fire in all my life." In the meantime, he and Denys buzzed the mountainside in search of the vehicle's occupants, who were in danger of being swept by the oncoming front. They called for a loudspeaker-equipped helicopter which alerted the hikers and allowed them to get off the mountain.

Even being airborne was hazardous in the turbulent conditions; winds were estimated to have reached more than 150 kilometres per hour just above the valley floor. When the fire jumped to Mount Ida and its lower flanks were just beginning to burn, someone suggested to Rapattacker Kevin Speilman that he should "jump in a helicopter and see if you can get the [radio] repeater off the top of the mountain before it burns." By the time he and his pilot were approaching the peak, Speilman said, "flames just curled over the top of the mountain. The trees were probably 30 metres high and the flames were at least twice that again." The repeater was quickly forgotten.

> Speilman: We were trying to get back to camp but the winds were incredible. This was just in a little [Bell] 206. We were flying into the wind and our air speed was 80 knots. Our ground speed was zero. The pilot had to really crank it up to get out us of there. When we got back to the camp we had to basically fly forward to try to land. We got our skids down but the camp was blowing away. Everything started blowing out of the storage tent—things flying at us—so we picked up and landed in a field off to the side.

Firecat pilot George Plawski flew one of the air tankers sent to Silver Creek on August 5. That blow-up, he commented, "was one that we shouldn't have been anywhere near."

> Plawski: It was huge. It was out of control. There was no visibility. The mechanical turbulence—turbulence caused by a combination of prevailing winds, fire-generated winds and local topography—was ridiculous. At one point I just started laughing. When you're doing full control deflections in both directions just to stay level, it becomes funny. Besides, we were doing absolutely no good whatsoever. With that much turbulence, hardly any of your load actually ends up on the ground. It was more likely to end up on the windshield of the guy behind you.

*Pilot George Plawski at the controls of a Conair Aviation DC-6. He flew a Firecat air tanker assigned to the Ash Wednesday firestorm. "It was huge. It was out of control. There was no visibility."* Photograph courtesy George Plawski

As devastating as the Silver Creek conflagration had become, it was only one of several wildfire emergencies simultaneously unleashed by the potent combination of drought, wind and lightning. While Silver Creek residents were rushing to save themselves and their possessions, people near Kamloops (the Greenstone fire), Vernon (the Aberdeen fire) and the west side of Okanagan Lake (the Fintry fire) were also caught up in fire-generated evacuations. Suppression resources were stretched to the breaking point, particularly in the case of aircraft. Dispatchers at the Kamloops Fire Centre spent hours making critical and often agonizing judgement calls—sending air tankers to protect homes described by ground forces

*A devastated Silver Creek neighbourhood. Embers rained down as flames leaped across the Salmon River Valley, creating a patchwork of destruction.*
James Murray photograph, *Salmon Arm Observer*

as being "extremely threatened," for example, and temporarily abandoning those merely "threatened" by advancing flames.

In some cases, firefighters were successful in saving endangered houses. In the case of 16 homes and a similar number of outbuildings they were not. There was often no apparent logic to the pattern of what burned and what didn't. As Speilman observed, "You could have a cedar shed right next to a house, and the shed would be fine but the house would be gone." In one instance he and his crew mates tore in their pickup down one long, narrow, tree-lined driveway as fire "like a jet" roared toward them through the trees.

> Speilman: We didn't have time to do anything but check the house for people, turn the truck around and lay rubber out of there. The fire was right there—up in the trees and moving. We ended up driving out with fire on both sides of

the road. And the funny thing about that house? It was fine. Everything around it burned, but it didn't get that key spark that burned long enough to do anything.

Similar scenarios played out around the valley throughout the night of August 5, a time of widespread panic. However, perhaps the most surprising aspect of the firestorm, beyond its sudden, overwhelming fury, was the fact that no one died in it. No one was seriously injured.

Farmers were among those who not only suffered the largest economic losses but also had the closest escapes in the name of protecting their livestock and property. Wendy Hawes had for several days watched the action being conducted in the hills above her Silver Creek farm and had already trailered most of her 15 horses to the safety of a friend's farm at the other end of the valley. Early in the afternoon of August 5 she and husband Dave took a break from setting up protective irrigation equipment around their place. Noticing that a windmill in her yard was spinning faster and more erratically than earlier in the day, Wendy evacuated her four remaining horses. She was driving back to her farm when the big wind struck. Flames, she saw, were running downhill toward the property of racehorse breeder Bucky Stockwell, who was away at the Kamloops racetrack. Taking charge, Hawes loaded a valuable stallion and an orphaned foal into her now-empty trailer. At that point Stockwell's wife arrived home and asked what would become of the 20 mares and foals pastured in a field directly in the fire's path. Hawes could think of only one solution, which was to drive the herd along the road to a set of corrals further down the valley. The problem, she knew, was that only the brood mares would be halter-broken. The others would be difficult to lead away from their home fields.

Hawes had no sooner coaxed the herd to the road when the animals tried to break back to the ranch. It was clear that the oncoming fire would give her time for only one more attempt.

> Hawes: This fireball was coming at us—I couldn't believe how fast it was coming. And so noisy. I've never lived through a tornado, but that's what I thought of. I could see this fire coming, tumbling like a rolling wave off the hill and down onto Bucky's farm.

She succeeded in haltering one mare and was relieved to see the remainder of the animals fall in line behind her. With the Silver Creek evacuation in high gear, traffic was now backing up on the road behind the horses. Hawes stopped a passing vehicle and asked the passenger in it to evacuate her truck and the two trailered horses, then she turned her attention back to the milling herd.

> Hawes: I didn't know whether the horse I had on a halter was broke or not and I was running out of wind from running. I thought I'd test the mare, so I held onto her mane real tight and just lifted my feet off the ground. She accepted that, so I jumped up on her back and rode out.

The herd had followed Hawes just a short distance down the road when the firestorm blew across the valley.

> Hawes: I looked over my shoulder and thought, holy man—we've got to get out of here. It happened in the snap of your fingers. If we hadn't got the horses out then, I think they would have been burned up, because everything on that farm burned except for the house—the firefighters saved that.
>
> There were so many things that were lucky. I was lucky to catch a broke horse. It was lucky I'd been there at that time; if I'd been there five minutes later it would have been too late. Some of the horses would probably have survived. The stallion wouldn't have—his barn and paddock burned right up. I was lucky that the horses all stayed with me. And I was lucky that I knew where there was a set of corrals.

After her five-kilometre ride, Hawes spent much of that night helping friends and neighbours evacuate livestock from the burning valley floor. During that time she learned that the firestorm had spared her home—though the roof had to be replaced owing to the number of ember holes burned in it—but that her barn, a hay shed containing 4,500 bales of winter feed, a tractor and various outbuildings had been destroyed. The hay shed was rebuilt by

*Riding instructor Wendy Hawes saved a neighbour's herd of horses by leading them out of the fire area. Hawes lost her own barn, equipment, hay and outbuildings to the inferno.* James Murray photograph, *Salmon Arm Observer*

Mennonite tradesmen who, in the style of the 1983 Swiss fire, were responsible for resurrecting many structures in the area. Mennonites also assisted in the August 5 evacuation and played a key role in setting up a food bank for dislocated families.

Two thousand head of livestock were evacuated from Salmon River Valley farms that afternoon. The only reported four-legged fatality occurred when, in the frenzied activity, a pig was run over—by a man named deBoer.

Once flames tore to the top of Mount Ida, a wind shift aimed them down the mountain's heavily timbered slopes toward the town of Salmon Arm. Another front ran north through the Fly Hills toward the hillside subdivision of Glen Eden, whose residents well knew the disastrous potential of wind-driven fire; 14 Glen Eden homes had burned in 1973 when wind blew a slash fire off a ridge and into that rural neighbourhood. Ironically, it was in a Glen Eden farmer's field that Roy Benson and his crews had established their fire camp on the 25th anniversary of that event.

As wind propelled flames toward the fire camp, Mike Worobey recalled, "It sounds a little bit corny, but it really seemed that the fire had a mind of its own." Worobey would not be the first firefighter to personify wildfire. Many people who have had close and prolonged contact with unrestrained flames have seen some form of sentience in what we objectively recognize as nothing more than combusting gases. "It was as though the main fire sent this other tongue of fire specifically to get the fire camp headquarters," Worobey said. Flames ran to within 40 metres of the fire camp, then died, sparing the huddle of tents and the now-evacuated subdivision. Worobey, having been raised in Salmon Arm, was now working on home ground—

literally. One of his last acts on the fire was to spray fire-retarding foam on his parents' Glen Eden house, the house in which he had been raised.

Over the following three days crews and equipment were reorganized and assigned to fight what had become a 6,300-hectare fire divided into three sectors: Silver Creek, Mount Ida and Glen Eden. Then, on August 9 firefighters received their most chilling forecast yet. As Benson remembered it, "Our weather prediction people came to us and said, 'The morning of August the tenth you're going to have winds as strong or stronger than on the fifth.'" Fire behaviour specialists briefly considered the effect that winds of the predicted strength would have on a smouldering giant whose perimeter now stretched for 58 kilometres. Benson had already worked the possibilities out for himself. An undramatic man, the fire boss described it as having the potential for "a massive disaster."

Given that the town of Salmon Arm would be located pretty much at the centre of such a disaster, Benson recommended a large-scale evacuation. Announced on the evening of August 9, the plan gave residents within the designated danger zone until seven o'clock the next morning to be out of town. That led to an exodus of 7,000 people, many of them taking refuge at evacuation centres established in Vernon, Sicamous and Kamloops. Adding to the sense of urgency was Premier Glen Clark's declaration of a state of emergency, the first in BC's history.

As residents of Salmon Arm hurriedly departed on the morning of Monday, August 10, a reverse flood of emergency response people took up positions in and around the town. Its overtaxed volunteer fire department brought in 65 fire trucks and 240 personnel from as far away as Golden, many of whom had just returned home after helping fight blazes sparked by the August 5 firestorm. Hundreds of emergency workers—police, municipal and wildland firefighters, Canadian Armed Forces personnel, ambulance attendants—tensed for an event of unimaginable ferocity.

Specific streets and avenues had been designated as fallback lines, should sections of Salmon Arm be sacrificed in an attempt to save others. "We had visions that, if the wind kept blowing, the fire was going to blow right into town," said Salmon Arm fire chief Ken Tebo. "We were ready but not totally optimistic." Helicopters were in the air early that morning, pounding 150 loads of retardant onto the critical north flank. Everyone held their breath and waited. And waited. "The forecast kept backing up," Benson said. "The wind would arrive by ten o'clock, by eleven, by noon, by four. By evening word finally came that the cold front had passed and we would not experience those winds." The winds did arrive but struck five kilometres south of Salmon Arm, flattening trees in the community of Deep Creek.

One Salmon Arm municipal official noted that area residents reacted to the fire in three ways: by refusing to talk about it, by taking pride in the community outpouring of support and generosity shown to victims, or by accusing the Forest Service of allowing the blaze to make its destructive sweep out of the Fly Hills. *Salmon Arm Observer* editor Heather Persson has written dozens of articles on the fire and its aftermath. Four years after the disaster, she noted that the events of August 1998 remain much discussed in the area generally, and are a point of considerable pain in the community of Silver Creek. "You touch that sore point and it comes pouring out—the anger and the blame."

*Many Silver Creek residents vented their anger at the Ministry of Forests, accusing the agency of incompetent firefighting.* James Murray photograph, *Salmon Arm Observer*

An investigation by Ombudsman Dulcie McCallum recommended 20 ways that the Ministry of Forests could improve its ability to fight fires and its communications with fire zone residents. But McCallum neither found fault with the province's response to the Silver Creek fire nor commented on possible compensation for losses resulting from it. However, the Ministry of Forests, aware that a class action suit is still being contemplated by Penticton homeowners burned out in the 1994 Garnet fire, knows that lawyers may yet face each other over its actions at Silver Creek. Directly criticizing the Ministry of Forests for its handling of the fire may have been outside McCallum's mandate, but it was a cause taken up passionately by victims and their supporters. On billboards, in letters to the editor and on radio and television the ministry was accused of not taking the fire seriously soon enough and of not fighting it effectively.

Considering the enormity of the Ash Wednesday event, the outpouring of anger was understandable. So was the rebuttal that a great deal of armchair firefighting took place during and after the fire. Jake Jacobson, a Salmon Arm fire warden with more than three decades of experience, began his involvement with the Silver Creek fire the evening it ignited; having "a bad feeling" about the blaze, he spent that night sitting in his truck in a Silver Creek driveway, monitoring the fire's activity. He ended his involvement when he helped extinguish the last burning snag three months later. He commented succinctly after one salvo in the barrage of criticism over the suppression effort.

*Salmon Arm fire warden Jake Jacobson worked on the Silver Creek fire from its first day to its last: "What you'd have up there now would be a bunch of grave markers because they'd all be dead."*
Keith Keller photograph

Jacobson: One person wrote to the newspaper here to say that the Forest Service should have stationed crews all over the Mount Ida side of the valley—that these crews could have jumped on the spot fires and put them out as they started, and that's how we would have saved the mountain. This person said he had a lot of firefighting experience. But he had no concept of the conditions at that time. This fire ran up Mount Ida at 90 metres a minute. Sure, we could have put crews on that hillside. And what we'd have up there now would be a whole lot of grave markers wherever those crews were because they'd all be dead.

# — A Red So Pure —

From the journal of Dave Jimmie, a member of a Sto:lo Nation unit crew nearly trapped in the 1998 Ash Wednesday blow-up.

A fire so wild and crazy that you could never imagine in your wildest dreams or nightmares the feeling of being surrounded by a red so pure it would make all colours look dull... it compares to the deepest blue in the furthest glacial lake. The kiss of death, so close that you can reach out and touch the harsh reality that is flaunted in your face. As the fury of the end came near and let us go, I realized that it was not our time, maybe we weren't meant to be here or maybe it was just bad timing... who knows? But some day it will come again; we will be faced with our fears, but maybe it will be different... maybe it will be water instead of fire.

# EPILOGUE:

# Is Black Bad?

It is an ironic, apparently heretical thought, but one that some people in the wildland fire business are thinking: at the right time, in the right place, maybe some smoke and flame is not such a bad thing after all. Not all of them have the conviction to go as far as Tom Lacey, a third-generation firefighter and the grandson of Edward Lacey of the 1931 McKinney firefight. Lacey grew up in the Interior dry belt, where he works today as the person responsible for wildfire control on one million hectares of crown real estate known as the Merritt fire zone. He suggests that one step in our necessary rethinking might be to revamp that universal symbol of fire suppression, the American bear known as Smokey. "I sometimes think," he says, "that we need to take that shovel away from Smokey the Bear and replace it with a drip torch."

Advocating for fire is a tough job in a culture with a suppression compulsion, though Lacey is not alone in believing in the cause. Stephen Pyne, an American who has fought wildland fire and studied it as an academic, asserts that obsessive fire suppression "has become a disordering process, and in much of the world it is the control of fire that is out of control." Fire is an integral part of forest ecosystems, a natural inhabitant of British Columbia's landscapes, though one more commonly found in some places than in others. Prior to European settlement, wildfire is thought to have annually burned 500,000 to 1 million hectares of land in what it now British Columbia. The average for the past five years is 26,000 hectares per year. In other words, we've eliminated fire from most of the terrain it once burned because we don't believe we can afford to have it there any more. For one thing, human settlement has expanded to the extent that the area where forests and human settlement meet—the "interface" in firefighting jargon or the "land-use omelette," as Pyne refers to it—has become huge, a central preoccupation for wildland firefighters in areas such as the Okanagan. For another thing, we'd much prefer to cut trees down than let them burn. As BC fire scientist Judi Beck puts it, "We're hoping to harvest what fire would naturally have harvested."

As has been the case when humans have attempted to eliminate or severely reduce animal populations—wolves, killer whales, birds of prey—for our own benefit, we're discovering that some ecosystems are impoverished by our actions. For example, those parklike bunchgrass and Ponderosa pine forests that typify BC's dry southern Interior are evolving into Douglas fir-dominated forests because fire no longer sweeps through them on a regular basis to clean out encroaching fir seedlings. Since, according to Lacey, most people "wouldn't know a natural forest if it fell on them," most people don't notice the transition or its consequences. However, that human-caused shift has a host of implications for indigenous plants and animals and for the increasing numbers of people who build homes in and around those forests. In the Interior dry belt centred in Kamloops, low-intensity fires once swept through on a five- to 15-year cycle, killing invasive plants and preventing ground-level fuels from accumulating to the point where they caused more intense, less suppressible conflagrations. "It's not a matter of whether fires are going to happen," Lacey notes, "it's a matter of when. All we're doing is delaying the inevitable. And we're to the point where it's scary. We're facing more hazardous fuel-loading by the year."

One antidote to that scare is what Lacey refers to as "pro-actively shedding the risk," which means informing people how to take responsibility for safeguarding their own homes and property against forest fires. Shedding the risk is something the Ministry of Forests' Protection program was seriously contemplating in the spring of 2002 in response to deep provincial budget cuts. How far that shedding would go had not been determined as fire season got underway, but it would possibly lead to wild-land firefighters taking less responsibility for protecting "improvements" in favour of protecting forests. This is an echo from the budget-cutting era of the 1931 McKinney fire, when burned-out ranchers were informed that Forest Branch resources "cannot in any sense be taken as a rural fire brigade for the protection of private property."

Two other responses to the threat from increased fuel-loading are "prescribed fire" and "prescribed wildfire." The former refers to intentional burning conducted under conditions chosen by the people who light them. At present, provincial wildlife officials annually burn thousands of hectares of forest to improve habitat. Most, though not all of that activity takes place in remote regions of the province; Lacey has helped co-ordinate burns designed to help wild sheep in the Okanagan and elk near Princeton. As an example of prescribed fire even closer to civilization he cites the 200-hectare hillside that he and the municipality of Merritt burned in three carefully calculated stages. That slope continued down into the town and ended in peoples' yards. Control lines were established at back fences and flames ran up from there, fireproofing that swath of forest for a half dozen or so years. He and his staff are preparing to burn another 800 to 1,000 hectares near the town's southern edge in the fall of 2002 or spring of 2003, again in the name of risk reduction.

Prescribed wildfire refers to the management of human-caused fires and those triggered by lightning. Most such blazes are clear threats and must be fought immediately. However, some can be assessed for possible beneficial impacts. Lacey notes the case of a 1998 fire near Spences Bridge that began with a farmer burning off bottomland to improve his fields. Flames escaped across a road and began running up a Douglas fir-Ponderosa pine hillside that naturally would have had about a 10-year fire interval but had been fire-free for 40 to 50 years. Lacey sized up the situation for risks and benefits and elected to let it burn. That's when his phone began ringing. "What the hell are you guys doing?" an irate resident inquired before providing his own answer: "You're not doing anything!" Lacey remembers his response. "I said to him, 'You know what? You're right.' I spent the next half hour on the phone with him and at the end of it he said, 'That makes sense'." The fire eventually burned through 150 hectares of crown land and Lacey was disappointed when rain put it out because he was hoping it would get considerably larger.

> Lacey: Normally people would have rushed out there with Cats and tank trucks and the expense and effort that goes along with it. Instead, we looked at it, monitored it, actually lit more fires around it at natural breaks, and used it as a positive rather than a negative event. We spent very little money on it, gained from the perspective of wildlife forage and cleaned up some of the fuels.

Lacey recognizes that he is considered something of a pyro-radical. As he says, "I've always been in the insane group on the left that's trying to go toward fire management rather than fire suppression." Among the people who look at him askance are some old-school firefighting colleagues who, he admits, "are having a real challenge" with his vision.

One person familiar with the management versus suppression debate as it relates to northern British Columbia is John Tigchelaar, the Ministry of Forests person responsible for fire protection in the Prince George region—the province's northern third. Tigchelaar shares Lacey's view that Smokey the Bear may be getting more good-guy credit than he deserves. Twenty years of flying over and directing firefighting in BC's north has convinced him of the gains to be had from letting certain fires go unfought or selectively fought, the reasons including insect control and the improvement of wildlife habitat.

There are other justifications for letting fires go. One is cost. Another is safety. Both issues came into play when a trio of lightning strikes ignited fires in the McGregor River area east of Prince George in 1998. An initial attack on the fires failed to put them out, and since this was taking place while much of the province's firefighting resources were being directed to emergencies at Kamloops and Salmon Arm, mounting a major campaign would have been logistically difficult. Furthermore, all three fires were burning on steep terrain which, combined with volatile burning conditions, made it risky to put people on the ground. Some good spruce-balsam forest was in danger of being burned, but Tigchelaar, having consulted with the local lands manager, finally elected to contain only one of those fires on one side in order to prevent it from entering a protected area. The result? About 1,000 hectares of forest burned, including some high quality timber. "Ten years ago," says Tigchelaar, "we probably would have fought those fires and put them out at great cost. I think we saved millions of dollars doing it the way we did, and I think we probably gained from the environmental perspective as well." As for economic justification in general, "If we're going to spend a million dollars, we'd better be able to justify saving a million and one. Otherwise, why do this? And that's not just timber values, that's all the values that have to be taken into consideration." When it comes to wilderness aesthetics, Tigchelaar says, people have to rethink the question: Is black bad?

In principle, letting fires burn is dead easy—you don't fight them. In practice, however, our culture has made not fighting fire as counter-intuitive as not saving a drowning person. Fighting a fire "is not a decision," Tigchelaar says. "It's like getting up in the morning and getting out of bed. We've been pounding into our staff for so many years: 'We've got to put 'em out, put 'em out, put 'em out.' It's a challenge now to get them to *not* put them out."

In an increasingly litigious age, the consequences of lighting prescribed fires or letting accidental fires burn are potentially huge. Tom Lacey acknowledges that his fire management ideas "make some managers shudder." What happened in New Mexico in May of 2000 didn't make things any easier for people who are inclined to light and manage wildfires. That was when a prescribed burn in a national park escaped and destroyed 200 homes in the town of Los Alamos. The disaster took place while BC's Ministry of Forests was experimenting with using prescribed fire to control a mountain pine beetle epidemic in Tweedsmuir Provincial Park, and its chilling effect—combined with disappointing results in terms of the number of insects killed—contributed to that experiment being shut down before it was attempted on a larger scale. When people know that intentional burning will take place, Lacey says, they at times demand a guarantee that the fire won't get away. The challenge he faces is that "if it isn't dry enough that it could get away, it probably isn't dry enough to meet the goals and objectives you've set out for your burn."

In the context of prescribed wildfire, "Nobody has been jumped on and criticized for deciding to fight a fire," says Tigchelaar. "But to make a decision *not* to fight one—that's much tougher. If you decide not to fight a fire you always get this 'What if?'—'What if it burns for 20 miles?'" In response, Tigchelaar argues that firefighters have at their disposal improved predictive powers made available through research into critical variables such as weather, moisture regimes and fire's behaviour in various fuel types.

> Tigchelaar: We're always talking about fire behaviour in the context of putting fires out and how difficult that's going to be. I think we have to start changing that. We need to say, "Let's not look at it from putting it out. Let's look at it from: What's this fire going to do? Where's it going to go? How big is this fire going to get and what should we be doing about it?" You can always turn around and say, "What if it burns for 20 miles?" Yes, it could, but what's the probability of that happening under the conditions that it's burning in? We need more help in finding answers to those questions.

In spite of his arguments in favour of letting selected fires run their courses, Tigchelaar acknowledges the anguish that can accompany such a determination. "The ones you let go are the ones you think about all the time—until it rains: 'God, did I make the right decision?'"

# Index